EAI/Springer Innovations in Communication and Computing

Series Editor

The impact of information technologies is creating a new world yet not fully understood. The extent and speed of economic, life style and social changes already perceived in everyday life is hard to estimate without understanding the technological driving forces behind it. This series presents contributed volumes featuring the latest research and development in the various information engineering technologies that play a key role in this process. The range of topics, focusing primarily on communications and computing engineering include, but are not limited to, wireless networks; mobile communication; design and learning; gaming; interaction; e-health and pervasive healthcare; energy management; smart grids; internet of things; cognitive radio networks; computation; cloud computing; ubiquitous connectivity, and in mode general smart living, smart cities, Internet of Things and more. The series publishes a combination of expanded papers selected from hosted and sponsored European Alliance for Innovation (EAI) conferences that present cutting edge, global research as well as provide new perspectives on traditional related engineering fields. This content, complemented with open calls for contribution of book titles and individual chapters, together maintain Springer's and EAI's high standards of academic excellence. The audience for the books consists of researchers, industry professionals, advanced level students as well as practitioners in related fields of activity include information and communication specialists, security experts, economists, urban planners, doctors, and in general representatives in all those walks of life affected ad contributing to the information revolution.

Indexing: This series is indexed in Scopus, Ei Compendex, and zbMATH.

About EAI - EAI is a grassroots member organization initiated through cooperation between businesses, public, private and government organizations to address the global challenges of Europe's future competitiveness and link the European Research community with its counterparts around the globe. EAI reaches out to hundreds of thousands of individual subscribers on all continents and collaborates with an institutional member base including Fortune 500 companies, government organizations, and educational institutions, provide a free research and innovation platform. Through its open free membership model EAI promotes a new research and innovation culture based on collaboration, connectivity and recognition of excellence by community.

Anandakumar Haldorai
Arulmurugan Ramu • Sudha Mohanram
Editors

6th EAI International Conference on Big Data Innovation for Sustainable Cognitive Computing

BDCC 2023

Springer

EAI
RESEARCH MEETS INNOVATION

Editors
Anandakumar Haldorai
Department of Computer Science and Engineering
Sri Eshwar College of Engineering
Coimbatore, Tamil Nadu, India

Arulmurugan Ramu
Department of Information Technology
Mattu University
Metu, Ethiopia

Sudha Mohanram
Sri Eshwar College of Engineering
Coimbatore, Tamil Nadu, India

International Conference on Big Data Innovation for Sustainable Cognitive Computing, https://bdcc-conf.eai-conferences.org/2023/

ISSN 2522-8595 ISSN 2522-8609 (electronic)
EAI/Springer Innovations in Communication and Computing
ISBN 978-3-031-54695-2 ISBN 978-3-031-54696-9 (eBook)
https://doi.org/10.1007/978-3-031-54696-9

This Springer imprint is published by the registered company Springer Nature Switzerland AG
The registered company address is: Gewerbestrasse 11, 6330 Cham, Switzerland

6th EAI International Conference on Big Data Innovation for Sustainable Cognitive Computing

BDCC 2023

- Contains proceedings from 6th EAI International Conference on Big Data Innovation for Sustainable Cognitive Computing (BDCC 2023), 15–16 December 2023, Coimbatore, India.
- Features topics ranging from big data analytics and artificial intelligence.
- Includes contributions from researchers, academics, and professionals from around the world.

This proceedings volume delves into the realm of big data innovation for sustainable cognitive computing. The featured papers provide in-depth insights into big data research, particularly focusing on the application of healthcare, finance, retail, and manufacturing. Many promising artificial intelligence techniques can be used for big data analytics, including representation learning, optimization, machine learning, deep learning, neural networks, and natural language processing. Big data and artificial intelligence produce exact and insightful decisions, predict future trends and behaviour, automate complex processes, and enhance customer engagement in various subfields, such as machine learning, natural language processing, computer vision, and robotics.

The methodologies of big data and cognitive computation play a crucial role in organizing, analyzing, collecting, and integrating data from a diverse range of sources, including sensors, agents, cloud devices, video/audio inputs, networks, log files, transactional applications, web platforms, and social media. This integration proves especially valuable for large organizations seeking comprehensive solutions for their data needs. The EAI International Conference on Big Data Innovation for Sustainable Cognitive Computing (BDCC 2023) provided a platform for the exchange of ideas and research findings in this dynamic field. Held on 15–16 December 2023 in Coimbatore, Tamil Nadu, India, the conference eased discussions and presentations on the latest advancements and applications in big data and cognitive computing.

Contents

About the Editors

Anandakumar Haldorai is a distinguished professional serving as a Professor and Research Head in the Department of Computer Science and Engineering at Sri Eshwar College of Engineering in Coimbatore, Tamil Nadu, India. He holds a Master's degree in Software Engineering from PSG College of Technology, Coimbatore, and earned his Ph.D. in Information and Communication Engineering from PSG College of Technology under Anna University, Chennai. Dr. Haldorai's extensive research expertise encompasses areas such as big data, cognitive radio networks, mobile communications, and networking protocols. With a remarkable academic and research record, he has authored over 176 research papers published in reputable international journals and presented at IEEE conferences. Additionally, he has authored 13 books and numerous book chapters with esteemed publishers including Springer and IGI. Dedicated to contributing to academic and professional communities, Dr. Haldorai holds editorial roles, including Editor-in-Chief for Keai—Elsevier—IJIN, Associate Editor for IEEE Access, and Area Editor for EAI Energy Web. He also serves as a reviewer for journals published by IEEE, IET, Springer, Inderscience, and Elsevier. Dr. Haldorai has been a guest editor for several journals affiliated with Elsevier, Springer, Inderscience, and others. His active involvement in various capacities, such as General Chair, Session Chair, and Panelist, reflects his commitment to academic events and conferences. Dr. Haldorai holds senior memberships with prominent professional organizations, including IEEE, IET, and ACM, and is recognized as a fellow member of the EAI research group. His multifaceted contributions showcase his leadership and commitment to advancing research and knowledge in the field of computer science and engineering.

Arulmurugan Ramu holds the position of Professor in the Department of Information Technology at Mattu University, Mattu, Ethiopia. He earned his B.Tech. in Information Technology, M.Tech. in Information Technology, and a Ph.D. in Information and Communication Engineering from Anna University. In his academic career, Dr. Ramu has imparted knowledge through teaching over 18 undergraduate and postgraduate courses, covering diverse subjects such as design thinking, image processing, pervasive computing, and artificial intelligence. His commitment to education and research is evident in the recognition he received, including the Best Young Faculty Award in 2018 and a nomination for the Best Young Researcher Award (Male). Dr. Ramu's research interests center on machine learning, deep learning, and bio-inspired optimizations, particularly as applied to image processing. He has contributed significantly to the academic literature, with authorship or co-authorship of more than 72 scientific papers and the publication of 6 patents. Beyond his research contributions, Dr. Ramu plays active roles as an editor, co-editor, or member of the editorial board for various scientific books, journals, and conferences. He is affiliated with professional organizations, including IEEE, IET, ACM, ISTE, and MICCAI, showcasing his engagement with the broader scientific community. Dr. Arulmurugan Ramu's diverse expertise and contributions reflect his dedication to advancing knowledge in computer science and engineering.

Sudha Mohanram is an accomplished academician with a diverse educational background. She graduated from the Government College of Engineering in Salem and earned her Master's in Engineering from the Coimbatore Institute of Technology. She furthered her academic journey by completing a Ph.D. in Electrical Engineering at Anna University, Chennai, in 2010. Dr. Mohanram started her teaching career as a Lecturer at the Government College of Technology in Coimbatore, amassing an impressive 22 years of teaching experience. A significant turning point in her career occurred around 13 years into teaching when her family established Sri Eshwar College of Engineering in 2008. Dr. Mohanram assumed the role of Secretary until 2011, after which she transitioned to become the Principal of the institution. Under her leadership, Sri Eshwar College of Engineering has flourished and gained prominence in Coimbatore, achieving notable success in academic excellence and placements within a short span. Dr. Mohanram's contributions to the institution's growth and success have been instrumental. In addition to her administrative responsibilities, Dr. Mohanram has actively contributed to research, publishing numerous papers in leading journals. Her affiliations include membership in professional organizations such as IEEE and the EAI research group, demonstrating her commitment to staying engaged with the broader academic and research community.

Part I
Big Data Analytics in Ubiquitous Data Science

Spontaneous Traffic Management System for Counting Vehicle Using IoT and Image Processing Techniques

R. Sreelatha, B. S. Mahalakshmi, G. V. Supriya, R. Sushmitha, and B. Chiranjeevi Nayak

1 Introduction

The Indian urban management system has various interconnected architectures, which are vital in regulating traffic. It is also considered one of the most crucial principles of the smart city. Rapid global progress is essential for the continued development of our planet. On the other hand, people do not have a smooth way to get around with modern transportation. Excessive traffic congestion compels frustrated and agitated drivers to postpone their commute to work or home, waste fuel, subject their cars to increased wear and tear, and may even incite incidents of road rage. Moreover, an expanding population immediately leads to a rise in issues related to traffic, i.e., excessive speed, collisions, hit-and-run incidents, and several others. Instances of criminal activities, such as the theft of mobile devices at traffic lights, occur often in urban areas, particularly during extended traffic congestion. Consequently, efficient traffic control has become an essential need for a thriving society. Currently, in most emerging countries, intelligent and flexible traffic control systems are favored over time-based methods. Most of the time, centralized systems and servers are used to keep an eye on this kind of traffic control. The Internet of Things (IoT), which has shown its use in almost all facets of our daily existence, may be seen as a mechanism for controlling traffic via a centralized server. For this work, the traffic flow control station may receive data on accounts of vehicles that travel through a particular road section located some distance before areas of heavy

R. Sreelatha (✉) · B. S. Mahalakshmi · G. V. Supriya · R. Sushmitha · B. Chiranjeevi Nayak
Department of Information Science and Engineering, BMS College of Engineering, Bangalore, Karnataka, India
e-mail: sree.ise@bmsce.ac.in; bsmahalakshmi.ise@bmsce.ac.in; supriya.is20@bmsce.ac.in; sushmitha.is20@bmsce.ac.in; chiranjeevi.is20@bmsce.ac.in

A. Haldorai et al. (eds.), *6th EAI International Conference on Big Data Innovation for Sustainable Cognitive Computing*, EAI/Springer Innovations in Communication and Computing, https://doi.org/10.1007/978-3-031-54696-9_1

traffic congestion. In order to control automobile access, the up-to-date information collected from the city's congested traffic intersection may be broadcast over the Internet and stored in the cloud. We will make use of OpenCV's image processing applications to count vehicles in real time. Our plan is being implemented at a low cost with minimal infrastructure required. The recommended plan effectively controls traffic on both centralized and local servers by using the principles of smart image processing. The first stage involves the acquisition of real-time video information based on the application of the recommended plan designed for urban India. Once the binary transformation and noise reduction are completed, the data is segmented into frames. Then, blob detection is used to identify objects of interest, and the proposed vehicle counting approach is utilized to estimate the number of cars. Real-time traffic control and management officials may also benefit from statistical representations of traffic data. It might also be vital for the planning of the future. The vehicle number is gathered via a real-time web-based database and is thereafter accessible to a controlling station. An advanced technological system that uses an amalgamation of cameras, sensors, and other Internet of Things (IoT) devices to accurately identify and quantify the volume of cars traversing a designated region. The system captures the data and sends it to the central control system for analysis. This data is then applied to enhance the flow of traffic, minimize congestion, and make insightful decisions concerning the development of infrastructure. The system can be incorporated with other intelligent city technologies like parking management models, traffic-related applications, and traffic lights projects, to enhance the overall convenience and efficiency of vehicle users. The system is able to facilitate real-time surveillance of traffic, predict the flow of traffic, and manage potential accidents. It can also provide essential data to city administrators to enable them to make data-driven decisions and enhance the overall traffic management of the urban environment.

In addition, the system not only provides applicable advantages but also assists in controlling severe environmental effects on traffic through a reduction of carbon gas emissions, so boosting the general health of residents.

In India's urban environment, the automated vehicle counting systems for IoT-assisted smart traffic control system may be considered as a smart development to the rapidly increasing traffic congestion and mobility issues faced by many towns in India. The growing number of vehicles on the road and the difficulties caused by traffic congestion in many Indian cities are the primary drivers behind the development of IoT-assisted smart traffic management systems that include automated vehicle counting systems. More people driving more cars means more congestion and longer commute times as a consequence of population growth, urbanization, and economic development. The environmental and economic impacts are substantial and also reduce the quality of life for locals. Also, manual counting is a major flaw in the old traffic control systems since it takes too much time and is easy for humans to make mistakes. With the increasing adoption of IoT technologies, automatic vehicle counting systems can provide real-time traffic data to help city officials make data-driven decisions and improve the overall management of traffic in the city.

In recent decades, the emphasis of several IoT applications has shifted towards smart city concepts due to the extensive urbanization. The use of resources and environmental issues like air and water pollution are just a few of the urban challenges. Keeping the concepts of traffic data and environmental parameters, it has become an important issue for understanding how cities affect the natural ecosystem and improving the effectiveness of everyday city activities like traffic control. Unwarranted traffic signals can have a negative effect on the efficiency and safety of traffic by presenting these issues: a prolonged delay; more traffic, more pollution in the air, and more gas use; disobedience to instructions; using less-than-ideal streets more frequently to avoid traffic signals; and an increase in the number of collisions, particularly rear-end collisions. By installing a signal, some types of crashes can be reduced in number or severity, while others will increase. The most common consequence of excessive signal use is an increase in total crashes, particularly rear-end collisions. Fixed-time signals operate in a predetermined order and always provide the same amounts of time for every traffic motion, regardless of whether traffic is present. The lights are changed by actuated signals in response to the traffic amounts in every direction. Before providing feedback regarding the availability of vehicles using another method, they use various sensors to effectively detect vehicles and alter the duration of green time to permit multiple vehicles through the intersection as possible.

2 Literature Survey

Because the proposed framework only makes use of hardware that is based on a Raspberry Pi, it appears to be a cost-effective and portable method for alleviating traffic congestion at peak intersections [1, 2]. Implementing a functioning monitoring system that accurately detects, tracks, and uses deep learning with convolutional neural networks is the ideal strategy [3]. The system's performance was evaluated against a variety of metrics, and an average accuracy of 86.56% was found. The Deep SORT and YOLOv3 algorithms were able to accurately count every kind of road object [4]. This chapter examines several intelligent traffic management systems. Green wave system, IR, and GSM were some of these [5]. TMS, or Traffic Monitoring Signal Timing, was created by utilizing a variety of IOT hardware component features.

The IOT platform is used to optimize traffic by efficiently allocating different times to each traffic signal taking into the consideration of number of vehicles in the road path [6]. The study demonstrated that image processing is a superior method for controlling the traffic light's state change. It showcases its ability to alleviate traffic congestion and eliminate time wastage resulting from a green signal on an unoccupied route. Since it employs real traffic images, it also detects vehicle presence more consistently [7]. The traditional traffic signal control system relies solely on the traffic police to regulate traffic based on traffic density. Numerous studies have been conducted on calculating the traffic density by controlling traffic light signals with sensors and image processing [8]. The proposed system performs well and contributes to lessening traffic congestion. The results show the current count of vehicles. The proposed system is good for managing vehicles in real time [9]. The timer display changes dynamically and the traffic density is calculated. This significant advantage prevents "unwanted waits" for vehicles in the more congested area [10]. The system is reliable and has a relatively low mean waiting time. Consequently, there is a reduction in both the fuel consumption and level of air pollution. It reduces traffic congestion and offers a less expensive and more practical means of modifying the existing traffic management system to make it more adaptive and intelligent [11]. The developed algorithm has counted 14,151 vehicles, in comparison to 14,601 accounts of real and visual data, and 450 vehicles have not been tallied; this demonstrates that the application method is 96.7% effective [12]. Our findings showed how the effectiveness of the neural network-based vehicle counting method is affected by various environmental conditions, camera field of views, and frame settings. By using an affordable embedded device such as the NVIDIA Jetson Nano Board, their system was able to get a 99% accuracy and recall rate for car counts via this investigation [13]. The experiment showed that the morning yielded the highest level of accuracy in counting, with a precision of 92.3%, while the afternoon showed the lowest level of accuracy, with a precision of 77.3% [14]. The test findings showcased the successful use of the displayed approach for vehicle recognition during nighttime, as well as counting. The technique attained a remarkable average rectification rate of 94% and a computation time of 44 frames per second [15].

The conclusion obtained from this study is that the performance of cross-correlation functions for the vehicle calculation system is measured by having a percentage of errors that reaches 10.09%.

3 Methodology Proposed

The automated counting of vehicles travelling through a selected location will be accomplished based on image processing approaches. The camera system is employed to record real-time videos of traffic flow along a specific route. After the video was recorded using a camera, the data was then processed using a mobile microprocessor processing device. After computing the mean number of vehicles per unit of time, the same data was then communicated to the central control system, and the identical processing system was used. The next two sub-sections go into great depth on how the work will be completed.

3.1 *Counting of Vehicle Using Image Processing Technique*

Data was collected from the camera system and then analyzed on a central processing unit (CPU) with the help of OpenCV software. Counting the occurrence of vehicles passing through a certain area requires a certain flow of images.

To begin, the first action is extracting each frame from the video stream.

- The primary image underwent Gaussian filtering to eliminate noise. In addition, a morphological closure procedure was used to accurately enclose the objects in the image.
- The image was converted into binary format by assigning a value of "1" or "0" to each pixel, depending on a predetermined threshold value. This threshold value was calculated using the OTSU bimodal threshold method. The binary threshold picture is created because it is simpler to discern object outlines in this type of image.

3.2 *Internet-Based Data Transmission for Traffic Management System*

There are four components to this Internet connection process between the end user computer system and the camera-driven processing system.

- Fetching rate per second Vehicle detection using Open CV information: After each predetermined amount of time, the mean value of the car that was present throughout the applicable time frame was obtained and recorded in a variable.
- Interfacing IoT cloud with microprocessor using Python: Secondly, a server and a mobile microprocessor system connected to the Internet are required for the remote transmission of processed data to a control room. We ran the server on the IoT cloud. Before we can utilize the IoT cloud server time database, we need to make sure it's compatible with the Controller or Processor. We will utilize Python to do this.
- Connecting with IoT (real time database): Upon completion of the integration process, the subsequent step included establishing a lint with our database. We have established a new database on the IoT cloud specifically for this objective and next, established a connection between Python and the database by using Credentials, which include API keys (Application Programming Interface) and the End Point URL (Universal Resource Locator).
- Sending vehicle count data remotely to real time cloud: Our objective was to transmit vehicle count data remotely to the server after establishing a connection with the database. We accomplished this by utilizing Python and the Time Stamp to transfer our Open CV vehicle count data to the IoT cloud system through the web. Once the data has been provided using the previously described techniques, the process is repeated once per second to continuously provide the capturing

system's monitoring information. The result of the work is completed and the creation of the full.

3.3 The Method of Vehicle Detection and Tracking

The input data for the vision system comprises sequential images obtained from a camera. The graphics portray the immediate ecosystem in front of the vehicle, including the road, adjacent vehicles, trees, and bridges. The main objective of the system is to differentiate automobiles from mobile and stationary entities in photographs and identify them as vehicles. This task is challenging due to the unpredictable terrain and the fluctuating lighting conditions caused by the time of day and weather. It is difficult to identify automobiles that approach the scene unexpectedly. Cars and vehicles of varying speeds, sizes, and looks emerge.

First, we will go over how passing cars are identified using motion information from numerous consecutive picture frames. Next, we explore the ability of an adaptive feature-based method to identify cars that are far away. These cars usually have less movement compared to the camera-assisted vehicle. Nevertheless, promptly recognizing automobiles based on only one or two photographs is difficult and can only be done accurately in favorable circumstances, such as when there is a significant difference in brightness between the vehicles and the background. Consequently, in cases when an object cannot be immediately identified, our system analyzes many picture frames and utilizes its tracking abilities to detect cars.

3.4 NodeMCU

One microcontroller made by Espressif Systems is known as the ESP8266. As a bridge between Wi-Fi and present-day microcontrollers, the ESP8266 is a self-sufficient Wi-Fi networking device. It is an executed self-contained application. The module contains the USB connection with a wide range of pin-outs. The NodeMCU devkit is attached to your laptop and acts as an Arduino using a micro-USB cord. It is also a breadboard-compatible (Fig. 1).

4 Experimental Results Discussion

This section describes the experiments that were carried out and the results that were achieved. CCTV cameras or conventional cameras can be used to capture live traffic video. Already existing footages of cities have been taken for the intended job. These movies accurately depict road and traffic situations in metropolitan locations. The video sequences show a peek of automotive traffic in urban areas. It is

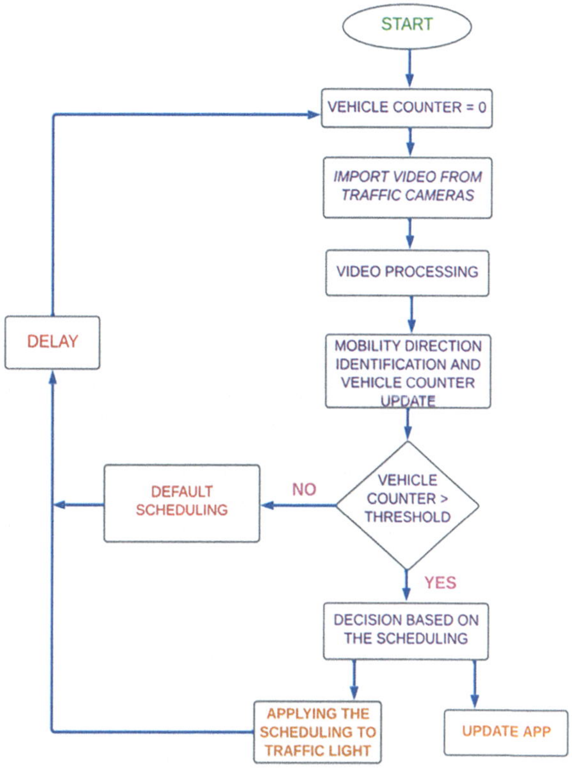

Fig. 1 Flow chart for vehicle counting and updating traffic signal

obvious from the preceding discussion that vehicle identification and counting are difficult tasks in these types of circumstances as a result of the availability of a high account of various types of vehicles and the lack of a systemized lane regulation method. As a result, before beginning object contour detection, the raw photos must be processed. The outcome of each phase is compared with the recommended model's execution, as detailed in methodology section of this work. The photos produced during the entire procedure are depicted in Fig. 2. The process begins by retrieving a single picture frame from a live CCTV video, as seen in Fig. 2a. The resulting foreground image obtained after applying Gaussian background reduction using OpenCV is shown in Fig. 2b. Noise present in picture's foreground was removed by subjecting it to a noise reduction filter. In addition, the morphological operations are employed to effectively enhance the item shapes, as shown in Fig. 2c. The crucial data is shown in the final diagram, namely, in Fig. 2d. The blue lines indicate the selected path for identifying the flow of vehicle counts. The contour lines within the specified area of interest are selected. Furthermore, once they achieve the size threshold level criterion, they are regarded as vehicles. As seen in the picture, two cars are identified on the specified road section and are determined

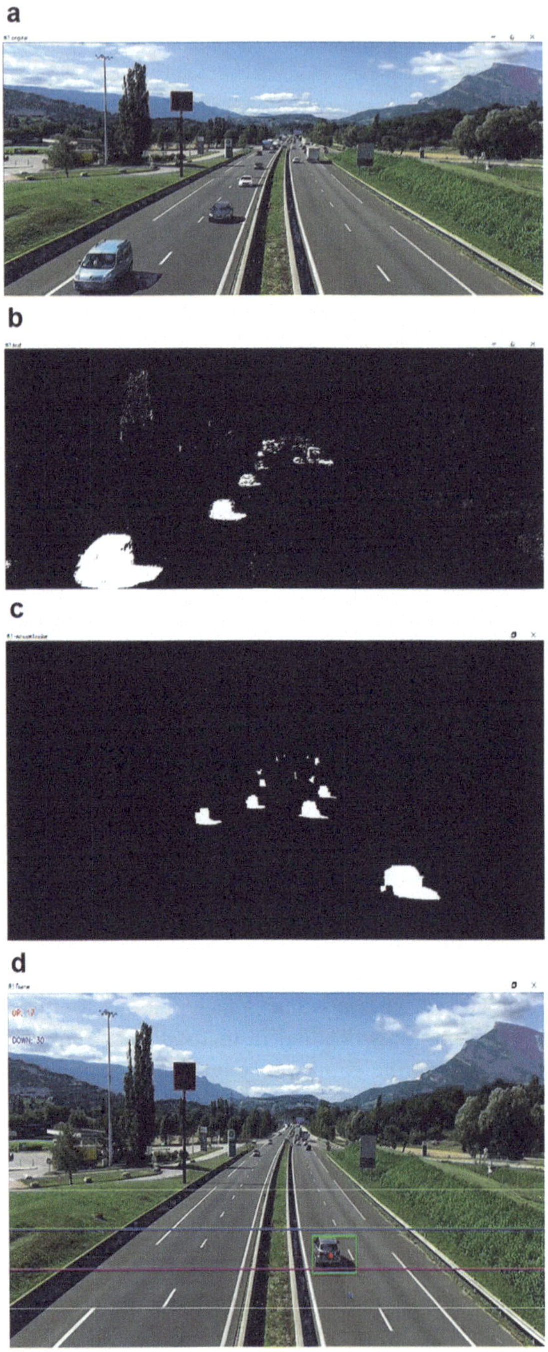

Fig. 2 (**a**) Initial video frame. (**b**) Foreground. (**c**) Image after morphological and noise reduction processes. (**d**) Image of vehicle detected by contour detection

Fig. 3 IoT LEDs

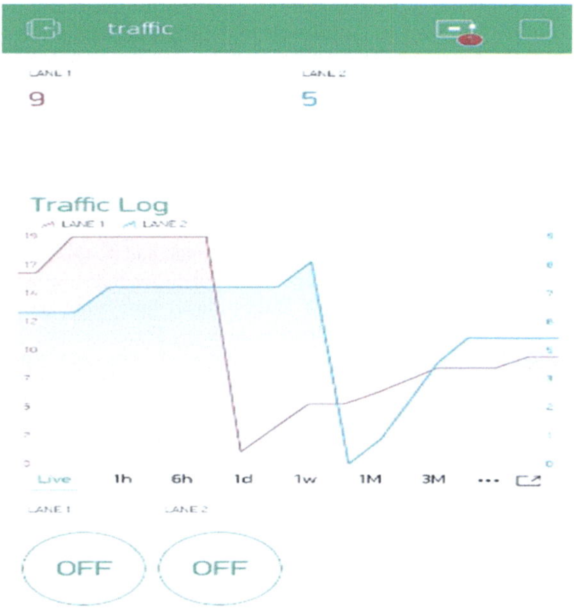

Fig. 4 Blynk app output

by shading green boxes around them. Figure 3 shows the output of traffic light scheduling using IoT. Upon computing the mean number of vehicles through a designated region of interest, this data is transferred to the central traffic control system over the Internet. Figure 4 illustrates the output data acquired from the remote-control module. The real-time Blynk database shown in the previous figure updates its value following the selected time interval. Consequently, data on the overall quantity of vehicles available at any certain moment is sent to the traffic control station. By using this information, traffic congestion at a crucial city junction may be averted when there is an unanticipated surge in traffic flow at certain areas beforehand. Traffic channels can be altered so that an excessive buildup of vehicles does not occur.

5 Conclusions

The computer vision technology along with Internet of Things (IOT) can be combined to design a traffic management system, which is artificially intelligent. The proposed framework makes use of hardware that is based on a Raspberry pi, which is a portable, cost-effective method for alleviating traffic congestion at peak intersections. The model was trained to correctly classify and recognize various objects from a video and image dataset under adverse weather conditions.

The vehicle counting system has many applications. In addition to leveraging information from vehicle counts, smart traffic lights may also be made. Depending on how many cars are at the signal, traffic light durations may be adjusted. Smart cities will find this beneficial. Using the Internet of Things, data on traffic congestion may now be gathered more rapidly and precisely.

The entire vehicle diagnosis process could be automated as part of the system's future development, allowing real-time classification information on various vehicle types to be provided so that more force can be diverted for traffic management.

References

1. S. Singh, B. Singh, Ramandeep, B. Singh and A. Das, "Automatic Vehicle Counting for IoT based Smart Traffic Management System for Indian urban Settings," 2019 4th International Conference on Internet of Things: Smart Innovation and Usages (IoT-SIU), 2019, pp. 1–6, https://doi.org/10.1109/IoT-SIU.2019.8777722.
2. G. M. Lingani, D. B. Rawat and M. Garuba, "Smart Traffic Management System using Deep Learning for Smart City Applications," 2019 IEEE 9th Annual Computing and Communication Workshop and Conference (CCWC), 2019, pp. 0101–0106, https://doi.org/10.1109/CCWC.2019.8666539.
3. R. Kejriwal, H. J. Ritika, A. Arora and Mohana, "Vehicle Detection and Counting using Deep Learning based YOLO and Deep SORT Algorithm for Urban Traffic Management System," 2022 First International Conference on Electrical, Electronics, Information and Communication Technologies (ICEEICT), 2022, pp. 1–6, https://doi.org/10.1109/ICEEICT53079.2022.9768653.
4. O. Avatefipour and F. Sadry, "Traffic Management System Using IoT Technology – A Comparative Review," 2018 IEEE International Conference on Electro/Information Technology (EIT), 2018, pp. 1041–1047, https://doi.org/10.1109/EIT.2018.8500246.
5. S.K Janahan, M.R.M. Veeramanickam, S. Arun, Kumar Narayanan, R. Anandan, Shaik Javed Parvez "IoT based smart traffic signal monitoring system using vehicles counts," 2018 IEEE 6th World Forum on Internet of Things (WF-IoT), 2018, pp. 1–6, https://doi.org/10.1109/WF-IoT48130.2018.9331084.
6. Mr. Sunil S. Sonawane, Bhagyashree Kenchannvar, Kajal Kumbhar, Pranjali Salunkhe, Smart Traffic Light Control Using Image Processing1 Lecturer, 2, 3, 4 Student, Department of Information Technology, AISSMS's Polytechnic, Pune, India ISSN: 2321-9653; IC Value: 45.98; SJ Impact Factor: 7.538 Volume 10 Issue III Mar 2022
7. Ashok, P.V, SivaSankari, S., Vignesh Mani, Suresh Sankaranarayanan, IoT Based Traffic Signalling System International Journal of Applied Engineering Research ISSN 0973-4562 Volume 12, Number 19 (2017) pp. 8264–8269

8. E. Basil and S. D. Sawant, "IoT based traffic light control system using Raspberry Pi," *2017 International Conference on Energy, Communication, Data Analytics and Soft Computing (ICECDS)*, 2017, pp. 1078–1081, https://doi.org/10.1109/ICECDS.2017.838960
9. L. P. J. Rani, M. K. Kumar, K. S. Naresh and S. Vignesh, "Dynamic traffic management system using infrared (IR) and Internet of Things (IoT)," 2017 Third International Conference on Science Technology Engineering & Management (ICONSTEM), 2017, pp. 353–357, https://doi.org/10.1109/ICONSTEM.2017.8261308.
10. Sakhare, Nitin Nandkumar et al. "Image Processing and IoT Based Dynamic Traffic Management System." *International journal of scientific research in science, engineering and technology* (2020): 180–188.
11. E. Chávez-Alarcón and A. Roman-Gonzalez, "Analysis of Vehicle Flow Using Auto Vehicle Counting," 2018 IEEE Third Ecuador Technical Chapters Meeting (ETCM), 2018, pp. 1–5, https://doi.org/10.1109/ETCM.2018.8580311.
12. M. Lücking, E. Rivera, L. Kohout, C. Zimmermann, D. Polad and W. Stork, "A video-based vehicle counting system using an embedded device in realistic traffic conditions," 2020 IEEE 6th World Forum on Internet of Things (WF-IoT), 2020, pp. 1–6, https://doi.org/10.1109/WF-IoT48130.2020.9221094.
13. H. Yuliandoko, M. D. Ayatullah and F. E. Purwita, "Automatic Vehicle Counting Using Raspberry pi and Background Subtractions Method in The Sidoarjo Toll Road," 2019 2nd International Conference of Computer and Informatics Engineering (IC2IE), 2019, pp. 83–86, https://doi.org/10.1109/IC2IE47452.2019.8940899.
14. Chih-Hsien Hsia, Shih-Chieh Yen, and Jing-Han Jang, An Intelligent IoT-based Vision System for Nighttime Vehicle Detection and Energy Saving – 2019
15. D. K. Soundra, M. Abdurohman and A. G. Putrada, "IoT-Based Road Vehicle Counter Using Ultrasound Sensor and Cross-Correlation Algorithm," 2020 8th International Conference on Information and Communication Technology (ICoICT), 2020, pp. 1–5, https://doi.org/10.1109/ICoICT49345.2020.9166350.

Quality of Service Protection in 5G Mobile Ad Hoc Networks with Behavior Learning-Enhanced CNN-AODV Routing

T. Saravanan, D. Venkatesh, and S. Deepa

1 Introduction

The technology that enables wireless communication is quickly advancing and assuming an increasingly significant role in the functioning of data communication networks. MANET, or portable randomly system, is a type of cell phone system types that is seeing increased use across a variety of industries, such as [1] smart cities, smart farms, smart traffic, and IoT ecosystems. To put it simply, a MANET is a distributed network in which no single entity has control. The topology is dynamic because of the constant motion of the nodes. Because of this, the routing table on each node needs to be regularly updated so that it can adapt to new topologies [2]. For this reason, MANETs require a reliable routing protocol.

The latest study and deployment of 5G-based MANET [3] aim to broaden MANET's application space within the context of growing 5G network technology. Like the 5G-based MANET depicted in Fig. 1, many MANET nodes communicate with 5G nodes to take advantage of the network's multimedia capabilities, as in [4]. 5G nodes, such as routers, access points, and mobile devices, can serve as gateways between MANET nodes and the 5G network base stations. Many distinctive features of 5G network technology stand out, but one is the widespread deployment of broadband transmission channels. This paves the way for 5G-based MANETs to have improved throughput and lower end-to-end delay. However, the 5G-based MANET faces a serious challenge in meeting the high traffic demand: The number of available network nodes is limited. This is a big difficulty for MANETs based on

T. Saravanan (✉) · D. Venkatesh
GITAM School of Technology, GITAM (Deemed to be University), Bangalore, India

S. Deepa
Government College of Engineering, Salem, India

A. Haldorai et al. (eds.), *6th EAI International Conference on Big Data Innovation for Sustainable Cognitive Computing*, EAI/Springer Innovations in Communication and Computing, https://doi.org/10.1007/978-3-031-54696-9_2

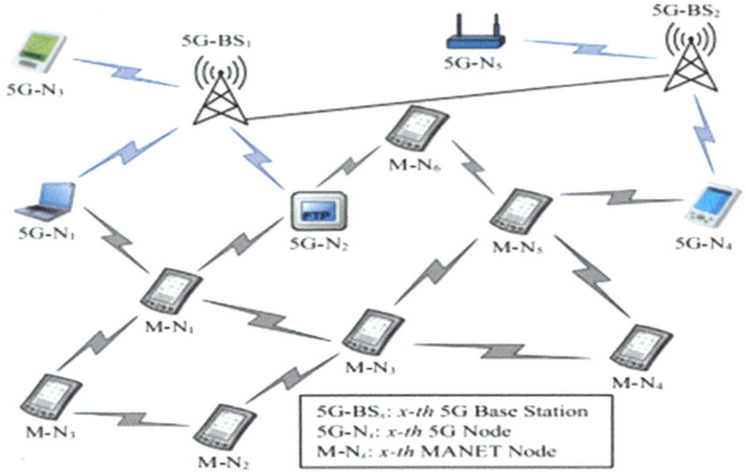

Fig. 1 Use of a 5G MANET in a typical situation

5G when it comes to preserving QoS. As a result, numerous academic institutions have recently focused on how to boost the efficiency of 5G-based MANET. In [5], the authors suggested a low-power routing protocol for 5G MANETs. A routing statistic that took into account both hop count and energy consumption was the idea behind their system. The routing algorithm will utilize this measure to choose a path. The authors used NS-2 simulation to show that, in comparison to more conventional routing protocols, their suggested approach significantly increases both lifetime and system performance. One of the most effective routing algorithms for 5G-based MANETs is proposed in [6], an improved fuzzy ant colony method for 5GMANET. The OFACA-5G algorithm is intended to safeguard MANETs.

The application of reinforcement learning (RL) in real-time scenarios within mobile ad hoc networks (MANETs) poses both challenges and opportunities. To improve the feasibility of RL in MANETs, research focuses on developing lightweight and distributed RL algorithms, efficient state representation techniques, and mechanisms to handle decentralized learning. Additionally [7], hybrid approaches that combine RL with traditional optimization methods or heuristics could provide more practical solutions for real-time applications within MANETs.

After routing is complete, it then utilizes the layout of a decentralized artificial intelligence recognition paradigm to discover and destroy rogue nodes. Within a millimeter wave mobile ad hoc system, the investigators of [8] examine the performance of well-known transit methods. Ad hoc on-demand distance vector, destination-sequenced distance vector, and optimal link state routing were among the many attempted sending methods. By simulating the connection using NS-3, we find the use of higher millimeter wavelength speeds improves the link's throughput. The authors of [9] looked into ways to boost AODV's performance on VANETs. Installing a static junction node at the roadside unit improves the delay from start to finish, data transfer proportion, and interaction among vehicles.

Addressing the scalability and adaptability challenges in QoS protection within 5G MANETs using behavior learning-enhanced CNN-AODV routing requires innovative solutions that strike a balance between efficiency, adaptability, and robustness. It [10] involves designing algorithms and protocols that can dynamically adjust to varying network conditions while efficiently managing resources to ensure consistent QoS guarantees across the network. Continual research and development in this area are essential to meet the evolving demands of 5G MANETs and their diverse applications.

The use of real-life approaches to in recent years and efforts to improve routing protocol performance in wireless networks have intensified [11]. The authors of [12] propose a method for locating itineraries with more residual energy in RL-powered mobile ad hoc networks operating over 5G. The primary goal of the method they suggest is to extend the practical lifespan of pathways without reducing the frequency with which they must be retransmitted. The energy efficient q-learning AODV protocol, defined by [13], is a transport mechanism for wireless ad hoc nets that leverages RL to improve network lifespan. The EQ-AODV approach collects network status information via Q-learning. After that, the routes will be revised to account for the fresh data. When considering packet delivery efficiency and power usage, the EQ-AODV is shown to perform better than the AODV-SARSA [14] in the simulated results.

2 Power Aware Routing in Mobile Ad Hoc Networks

Energy-aware route discovery can be carried out in two distinct ways by maximum lifetime-routing protocols: [15] either globally or locally. Related research on simulating routing behavior in ad hoc networks is summarized here. The majority of these works combine established routing algorithms with established learning approaches to create intelligent routing.

All mobile nodes in Global Routing contribute to the process of route discovery by relaying RREQ (Route Request) packets. After that, the nodes at either end of the found pathways evaluate them using an energy-aware measure.

The idea is introduced in [16] through the time delay route request that each node sends. The time that a node keeps the RREQ packet is inversely proportionate to the power that it has left in its batteries. Therefore, routes with low-energy nodes are unlikely to be selected.

Maximum node lifetime with little energy usage is the goal of [17]. The first-visit ONMC RL algorithm is executed by each source node to determine the optimal path based on three primary criteria: minimum-energy path, maximum-minority residual battery path, and minimum-cost path.

Authors in [18] also opt to merge the routing protocol with the RL algorithm, similar to the work in [19]. They started by framing the issue as a sequential decision-making problem and then demonstrated how to transform routing into a reinforcement learning problem with a Markov decision process that is only partially observable.

[20] Here, we introduce the Energy-Aware span routing Protocol (EASRP), a novel algorithm that makes use of energy-efficient techniques like Span and the Adaptive Fidelity Energy Conservation Algorithm (AFECA) [21]. In order to further reduce power consumption, the Remote Activated Switch (RAS) hardware circuit is used to reawaken dormant nodes. These methods of conserving energy have a long history of use in reactive procedures. Using EASRP in a hybrid protocol, particularly a proactive one, does present certain challenges, though.

3 Our Proposed Algorithm

By considering real-life information like congestion, forwarding is a mechanism that is a particularly efficient approach that involves significant thinking and prompt action. On top of that, real life has been utilized for transportation in 5G-based MANETs. However, our implementation of RL is unique when compared to the literature. The following are some instances of cutting-edge developments that showcase this specificity.

The primary objective of proposing a modification to the CNN-AODV (Convolutional Neural Network-Ad Hoc On-Demand Distance Vector) protocol for 5G Mobile Ad Hoc Networks (MANETs) typically revolves around improving the network performance, efficiency, reliability, or addressing specific challenges associated with MANETs in the framework of 5G technology.

The intermediate nodes in a 5G-based MANET or other multi-hop wireless communication network, the signal-to-noise ratio of a given path is affected by augment and either interpret and forward. For the route rs, the distance d can be measured in the SNR at the last hop be denoted by (r) s,d.

$$\rho\max\left(s,d\right) = \max \forall i \in rs, d\left(\rho i\right) \tag{1}$$

where SNR of the hop hi,j is denoted by (kh) i,j.

To ensure that the receiving nodes receive the most recent information feasible, RL is utilized to update the nodes' max (s, d) and r (s, d) metrics throughout data transmission. Figure 2 depicts a sample SIDB for a unit in a 5G-based MANET, which is although the characters on the Internet connections signify the link's strength (in dB). The node's statistics indicate the amount of data messages waiting to be processed.

In Fig. 3, we see the steps required to update the SIDB and find a new route starting at node I, outlining the basic architecture of the RLI-AODV method. Node I's current neighbors are the letters J, K, L, and S. After receiving an ACK packet from an adjacent device, the Q-Learning is method modifies the SIDB. It uses the data in its SIDB to decode RREQ packets it receives in order to find new routes.

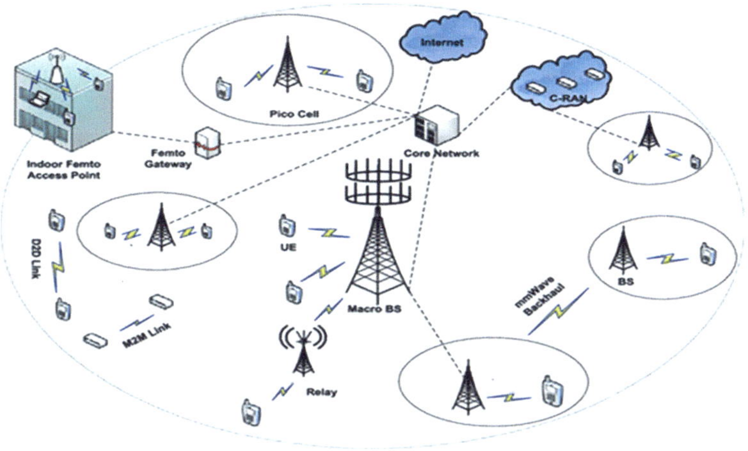

Fig. 2 5G-based MANET runs its own implementation with internal intelligence repository of each node

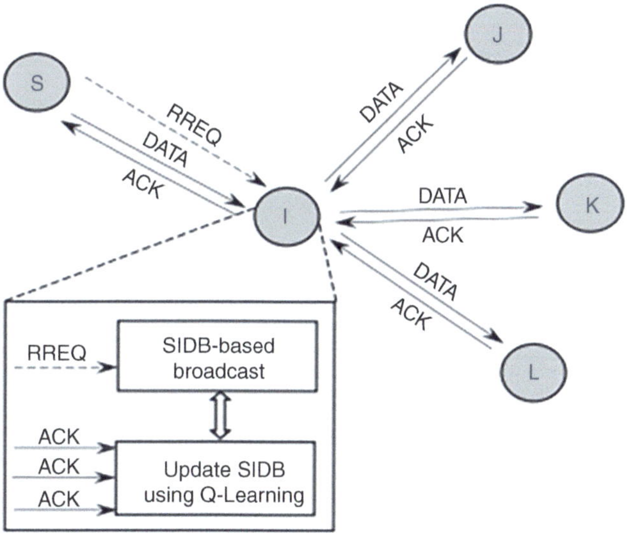

Fig. 3 Flow process RLI-AODV's system model for routing

SIDB will be adjusted through RL; the essence of RL is that it is a form of machine learning that is based on trial and error, which means that the design entails repeating every experiment but acquiring information from each one. An intermediary is responsible for on the role of a learner by interacting with its surroundings to determine the course of action that will yield the greatest number of rewards.

The value Q(s, a) represents the reward for performing action a while in state s, and it is calculated by an RL method called Q-Learning. The Q-value, often known as Q(s, a), can be computed by using the formula (2):

$$\mathrm{QoS}(\mathrm{si,aj}) = (1-\alpha)\mathrm{Qos}(\mathrm{si,a}) + \alpha\left[\mathrm{YR}(\mathrm{si,aj}) + \gamma \max \forall \mathrm{ia}' \in \mathrm{AQ}(\mathrm{si}', \mathrm{aj}')\right] \tag{2}$$

where the discount factor $\gamma \in [0, 1]$ and the learning rate $\alpha[0, 1]$ both play important roles. R(s, a) represents the payoff for doing an action in state s, with state s′ being the consequence of a. Applying Rr for routing methods, the triad "state, action, reward" [22] represents the action of identifying the way and modifying the system's status. To ensure that all 5G-based MANET nodes are always up-to-date on their status information, and our proposed algorithm employs RL. Here's how we may define the state-action.

3.1 Reward Triad

When considering the procedure of refreshing the structured integrated database and developing a new path through node

$$R\rho(\mathrm{s,aj,j}) = \max \forall \mathrm{i} \in \mathrm{rs,d}\left(\rho i, \rho \max(\mathrm{j,d})\right) \tag{3}$$

$$R\rho = \sqrt[n]{JR\beta(\mathrm{sk,ai,j})^n} + \begin{cases} \mathrm{ks,ai}, & n\,\mathrm{odd} \\ |\mathrm{s,aki}|, & n\,\mathrm{even} \end{cases} \tag{4}$$

The architecture of the RLI-AODV method is represented in Fig. 3. J, K, L, and S are at present siblings of node I. The Q-Learning technique updates its SIDB every time it receives an ACK packet from a neighbor during data transfer. I use the data in its SIDB to decode RREQ packets it receives to find new routes. The original AODV protocol didn't have this, though. We provide the specifics of how to handle RREQ packets.

3.2 State

Each possible state s is represented by the pair (traffic loads at each network node) P, B. Here, P = 1, 2, 3, …, n. Node i's traffic load, as indicated by the size of its packet queue, is represented by the set of values j in the set of packet queue sizes P. Sensitivity for a wireless channel li j is represented by Jhi(i, j) B, in which L is the set of all wireless links in the network.

3.3 Linking the Moving Stage

In every instance, a node in the network will take the first step by transmission of information to a subsequent network node partner, which will then relay the packet to its destination.

These ACK packets are used for communication between the nodes. Two one-byte fields, representing traffic load and signal-to-noise ratio, are appended to the ACK packet. Algorithm 1:

1: Use recurrent neural networks to adjust the volume of traffic passing via node I.
2: max (i, d) 0;
3: r (i, d) 0x F F F F;
4: end for
5: for (every time node I sends a data packet to node D) do // I Node
6: I's route cache must be consulted to determine which node J will receive the data packet.
7: Forward information to J.
8: if (data packet is delivered to J) then // at the J node
9: Make a package that acknowledges I as a reply; Obtain the values for max (j, d) and h(j, d) from J's SIDB and save them in the ACK packet;
11: Have it been an ACK packet to I. When the ACK is sent and received at the I node, condition 12 is met.
13: Examine the ACK packet for the values of max (j, d) and h(j, d);
14: Determine Q (s, ai,j) and Q(s, ai,j) using equations (6) and (7);
15: Modify node I's SIDB by writing: max (i, d) > Q(s, ai, j) and h(i, d) Q(s, ai, j);
16: End if
17: End if
18: End for

Each action yields a reward based on two measures: the greatest possible level of traffic encountered by the sending node and the signal-to-noise ratio (SNR) between the sending and receiving nodes

$$R\rho(\mathrm{s,ai,j}) = \max \forall \mathrm{i} \in \mathrm{rs,d}\left(\rho i, \rho \max(\mathrm{j,d})\right) \tag{5}$$

$$R\beta(\mathrm{s,ai,j}) = \mathrm{aj{\cdot}ai}\,\psi + \frac{\partial^2\psi}{\partial x^2} + \frac{\partial^2\psi}{\partial y^2} + \frac{\partial^2\psi}{\partial z^2} + \frac{1}{r^2\sin\theta} \left[\sin\theta\frac{\partial}{\partial r}\left(r^2\frac{\partial\psi}{\partial r}\right) + \frac{\partial}{\partial\theta}\left(\sin\theta\frac{\partial\psi}{\partial\theta}\right) + \frac{1}{\sin\theta}\frac{\partial^2\psi}{\partial\varphi^2}\right] \tag{6}$$

where s represents the current state and ai,j represents the action in which node i transmits a packet to neighbor node j. The values of R (s, ai,j) and R(s, ai,j) are determined by the formulas (1) and (2), respectively.

The suggested solution employs the Q-learning algorithm at each node we have developed in order to routinely update the aforementioned two metrics. The following change has been made to the Q-value update equation:

$$Q\rho(\text{s,ai,j}) = (1-\alpha)Q\rho(\text{s,ai,j}) + \alpha\left[R\rho(\text{s,ai,j}) + \gamma\max \forall \text{a'} \in \text{A}Q\rho(\text{s',a'i,j})\right] \quad (7)$$

$$Q\beta(\text{s,ai,j}) = (1-\alpha)Q\beta(\text{s,ai,j}) + \alpha\left[R\beta(\text{s,ai,j}) + \gamma\max \forall \text{a'} \in \text{A}Q\beta(\text{s',a'i,j})\right] \quad (8)$$

where si represents the current energy state and si′ represents the state of affairs following the execution of ai,j, is the learning rate, and is a discount factor in the range [0, 1]. The maximum traffic load Q-values and SNR from the initial node to the goal are Qρ(s,ai,j) and Qβ(s,ai,j), respectively. Pseudo code for the RL-based algorithm that updates traffic loads at individual nodes is presented in Algorithm 1.

Algorithm RLI-AODV for Rapidly Changing Environments
Our suggested method, RLI-AODV, is represented in pseudo code as Algorithms 2 and 3, and it takes advantage of the updated route discovered by the AODV protocol. Route discovery in RLI-AODV varies from that in AODV principally because traffic load and SNR restrictions are instructions. As a consequence, the newly established routes are confined to travelling through high-density traffic nodes, and there is no guarantee that they will connect with the SNR.

The following parameters can be measured: signal-to-noise-and-interference percentage, speed, and latency from beginning to end measured in a simulated network to assess RLI-AODV's performance in comparison to AODV via OMNeT++ 7.0 simulation [23] as well as the INET Platform 4.0 [24]. The initial conditions for the simulation are listed in Table 1. Each simulation scenario is executed ten times because station migration and transport growth are inherently erratic. In this study, we present the mean of ten independent experiments. The bandwidth of the network

Table 1 Simulation metrics values mobility model

Simulation metrics	Values
MANET deployed area	600 m × 600 m
Mobility model	Random way point
Transmission communication range	300 m
Number of mobile nodes	10, 20, 30, 40, 50, 60, 70, 80, 90, 100
Data packets count	10, 20, 30, 40, 50, 60, 70, 80, 90,100
Speed of nodes	2–10 m/s
Simulation time	100 s
Protocol	CNN-RLI-AODV, -RLI -AOMDV
Pause time	10 s
Number of runs	64

is the first thing to investigate typical throughput against simulation time for a varying number of nodes. The RLI-AODV algorithm is clearly able to produce greater throughput than the AODV method.

Second algorithm, RLI-AODV, is used at the S source node.

1: S reads the SIDB entries max (s, d) and r (s, d);
2: Create RREQ packet; send it out to all of S's neighbors;
3: while (twait timeout) do
4: if (RREP packet arrives), then
5: receive RREP and update new route into S's route cache;
6: end if;
7: end while;
10: else
11: Don't bother trying to plot an alternate path;
12: Give up if

The third RLI-AODV algorithm operates at the final or intermediate nodes (I).

1: RREQ packet arrives at Node I;
2: if (This RREQ has been received before), then toss the packet;
3: Toss the RREQ packet and call it a day;
4: otherwise, go to Node II.
5: If (I is an intermediate node), then
6: Consult the SIDB for the values of max (i, d) and r(i, d);
7: If (max (i, d) max && r(i, d) min), then
8: if (there is the fresh route ri,d in RC of I) &&
10: (DSN of ri,d > DSN in RREQ) then
11: join routes rs,i with ri,d;
12: create RREP packet and send to S via reverse route;
13: else
14: broadcast RREQ to all neighbors of I;
15: end if
16: else
17: broadcast RREQ to all neighbors of I;
18: else
19: Toss the RREQ package at the end (17), and then stop if (18)
19: if not // I am the final node
20: Modify the backwards path from S into I's RC;
21: Make an RREP package.
22: Employ a reversal of forwarding to transmit RREP to the route;
23: end if
24: end if

4 Experimental Analysis

Suppose the nodes in a network move at a usual, the AODV typically has a throughput of 6 (m/s) and other techniques are 172.65 and 256.64 (kbps), respectively. As a result, the RLIAODV technique boosts throughput by 64.56 kbps. Figure 5 shows how the RLI-AODV algorithm boosts throughput by an average of 66.12 kbps with nodes moving at a mean speed of 16 (m/s). RLI-AODV always results in higher throughput, whether the nodes move at 5 or 30 m/s. The data in Fig. 5 demonstrates that the network's EED is then examined and significant for 5G mobile networks. Figure 4 depicts the combined performance comparison between RLI-CNN_AODV and CNN-AODV for a 64-node circuit moving at 30 (m/s). Compared to the AODV, for method, the EED associated with the RLI-AODV method is generally less. An AODV AEDV is typically 1.281 (s), while an RLIAODV AEDV is 2.153 (s). It happens because RLI-AODV analyzes traffic flow at the intermediary nodes during path identification. This minimizes EED by minimizing the amount of time spent waiting in lines. Figure 6 depicts the average EED difference between the two approaches for rates of node motion that vary from 6 to 36 (m/s). RLI-AODV is a method that outperforms the AODV technique in all scenarios.

Figure 4 illustrates that RLI-CNN_AODV achieves higher throughput of 100%, while others achieve only like 70%, 47%, 20%, and 4%.

Figure 5 illustrates that RLI-CNN_AODV achieves higher throughput of 20%, while others achieve only like 8%, 5%, and 3%.

At the end, we look at the SNIR at each and every node in the network. As illustrated in Fig. 6, the lowest the relationship between SNIR over every junction and changes in average mobility speed has been studied. As observed, the RLI-AODV method typically achieves a higher SNIR compared to the AODV technique. This is because the SNIR constraint has been taken into account by the RLI-AODV algorithm during the route discovery phase.

Fig. 4 Simulation delay with packet size flow operation

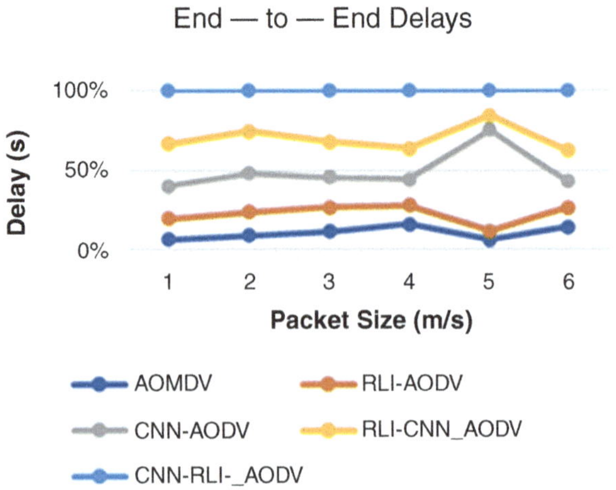

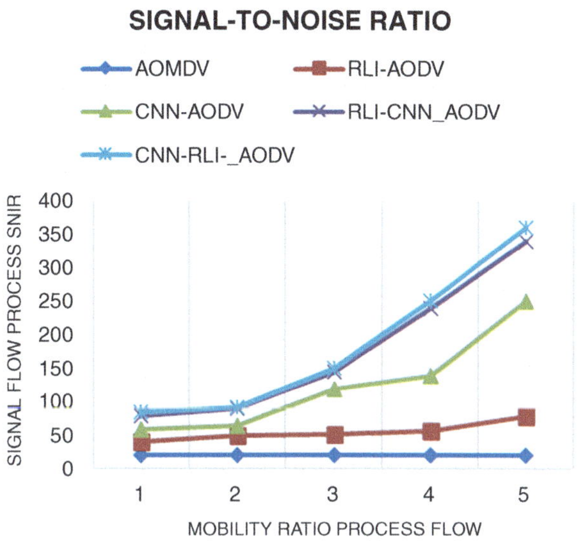

Fig. 5 Simulation throughput for mobile displacement flow operation RLI-CNN_AODV

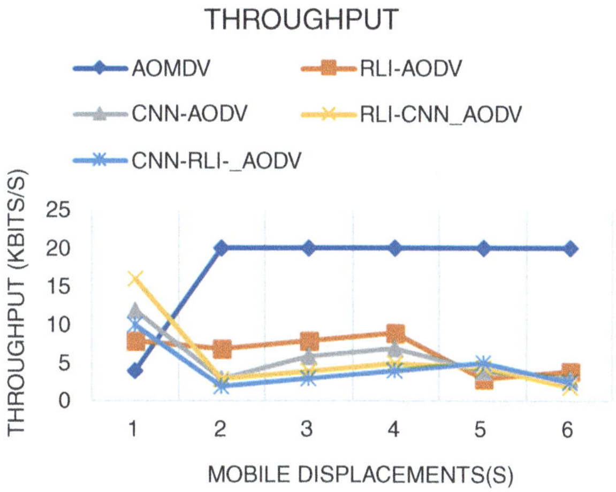

Fig. 6 Simulation signal to noise ratio for RLI-CNN_AODV operation

5 Conclusion

MANETs that are based on 5G are becoming increasingly common in the next generation of ad hoc wireless networking technologies. The fact that nodes in 5G networks are always coping with enormous traffic loads is the defining attribute of these networks. Although the very strict requirements put on end-to-end delay, probability of congestion, and other performance considerations, satisfactory communication has to be kept throughout that period for change. When applied to a MANET based on 5G network information technology, it becomes evident how insufficient the current MANET routing protocols are for 5G-based MANETs; we introduce a revised ad hoc on-demand distance vector (RLAODV) protocol. RL is used by every node to store data on traffic volumes and SNIR readings taken at intermediate nodes along routes. With the use of this database, we can determine the traffic density at each individual node. To guarantee service levels along a newly discovered route, the routing algorithm consults this database. Throughput, latency, and SNIR simulations showed that our proposed method was quite effective. We will keep working on the algorithm while taking other parameters into account in order to boost the performance of 5G-based MANET.

References

1. K.Q. Vu, V.K. Solanki, A.N. Le, A saving energy MANET routing protocol in 5G, in: S. Velliangiri, M. Gunasekaran, P. Karthikeyan (Eds.), Secure Communication for 5G and IoT Networks, Springer International Publishing, Cham, 2022, pp. 213–220.
2. S.K. Sarkar, T.G. Basavaraju, C. Puttamadappa, Ad Hoc Mobile Wireless Networks – Principles, Protocols, and Applications, Taylor & Francis Group, LLC, 2008.
3. R. Bharathy, T. Manikandan, P. Keerthivasan, P.M. Fayaz, A.M.S. Zafer, K.M. Krishnan, Typical MANET design for 5G communication network, in: Ambient Communications and Computer Systems, Springer Nature Singapore, Singapore, 2022, pp. 383–390.
4. F. Khan, K.-L. Yau, M. Ling, M. Imran, Y.-W. Chong; An intelligent cluster-based routing scheme in 5G flying ad hoc networks, Appl. Sci. 12 (2022) 3665.
5. Q. Vu Khanh, N. Ban, N. Han, An advanced energy efficient and high performance routing protocol for MANET in 5G, J. Commun. 13 (2018) 743–749.
6. M.F. Khan, K.-L.A. Yau, Route selection in 5G-based flying ad-hoc networks using reinforcement learning, in: Int. Conf. on Control Sys., Comp. and Eng., ICCSCE, 2020, pp. 23–28.
7. S. Sengan, et al., An Optimized Fuzzy Based Ant Colony Algorithm for 5G-MANET, Cmc -Tech. Sci. Press- 70 (2021) 1069–1087.
8. S.A. Alghamdi, Stable zone-based 5G clustered MANET using interest-region-based routing and gateway selection, Peer-to-Peer Netw. Appl. 14 (2021) 3559–3577.
9. M. Aljumaily, Routing protocols performance in mobile ad-hoc networks using millimeter wave, Int J Comput Netw Commun 10 (2018) 23–36.
10. Sahabuddin, A. Achmad, S. Syarif, Optimization of the 5G VANET routing protocol on AODV communication with static intersection node, in: 2021 3rd East Indonesia Conference on Computer and Information Technology, EIConCIT, 2021, pp. 171–176.
11. Y. Wang, Y. Tang, BRLR: A routing strategy for MANET based on reinforcement learning, in: 2021 IEEE 20th International Conference on Trust, Security and Privacy in Computing and Communications, TrustCom, 2021, pp. 1412–1417.
12. V. Duong Thi Thuy, L. Binh, IRSML: An intelligent routing algorithm based on machine learning in software defined wireless networking, ETRI J. 44 (2022).
13. M. Redha, S. Chikhi, Reinforcement Learning Based Routing Protocols Analysis for Mobile Ad-Hoc Networks, 2019, pp. 247–256.
14. T.-V.T. Duong, L.H. Binh, V.M. Ngo, Reinforcement learning for QoS guaranteed intelligent routing in Wireless Mesh Networks with heavy traffic load, ICT Express 8 (1) (2022) 18–24.
15. R. Mili, S. Chikhi, Reinforcement learning based routing protocols analysis for mobile ad-hoc networks, in: E. Renault, P. Mühlethaler, S. Boumerdassi (Eds.), Machine Learning for Networking, Springer, 2019, pp. 247–256.
16. S. Chettibi, S. Chikhi; An adaptive energy-aware routing protocol for MANETs using the SARSA reinforcement learning algorithm, in: 2012 IEEE Conference on Evolving and Adaptive Intelligent Systems, 2012, pp. 84–89.

17. H.A.A. Al-Rawi, M.A. Ng., K.-L.A. Yau, Application of reinforcement learning to routing in distributed wireless networks: A review, Artif. Intell. Rev. 43 (2015) 381–416.
18. A. Varga, OpenSim Ltd., OMNeT++ Simulation Manual, Version 6.x, 2022, [Online]. Available: https://omnetpp.org.
19. A. Virdis, M. Kirsche, Recent Advances in Network Simulation – The OMNeT++ Environmentand Its Ecosystem, Springer Nature Switzerland AG, 2019.
20. Kampitaki, D. G., & Economides, A. A. (2023). Selfishness in Mobile Ad-hoc Networks: A Literature Review on Detection Techniques and Prevention Mechanisms. IEEE Access.
21. Zabeeulla, M., Sharma, S. K., & Chauhan, S. P. S. (2023). Design and Modelling of hybrid network security method for increasing security in vehicular ad-hoc network. Measurement: Sensors, 29, 100878.
22. Goyal, P., Rishiwal, V., & Negi, A. (2023). A comprehensive survey on QoS for video transmission in heterogeneous mobile ad hoc network. Transactions on Emerging Telecommunications Technologies, e4775.
23. Krishnasamy, B., Muthaiah, L., Kamali Pushparaj, J. E., & Pandey, P. S. (2023). DIWGAN optimized with Namib Beetle Optimization Algorithm for intrusion detection in mobile ad hoc networks. IETE Journal of Research, 1-20.
24. Lakshmi, G. V., & Vaishnavi, P. (2023). A trusted security approach to detect and isolate routing attacks in mobile ad hoc networks. Journal of Engineering Research.

Decoding Li-Ion Battery Longevity: Domain-Driven Approach with Advanced Feature Engineering and Machine Learning for RUL and SOH Estimation

Abirlal Metya, Mohammad Shadan, and Garlapati Anusha

1 Introduction

Electric vehicles (EVs), micro-grids, consumer electronics, and renewable energy sources are just a few of the numerous modern-day products and applications that depend on lithium ion (Li-Ion) batteries for energy storage. These batteries are vital in reducing carbon emissions in the global economy. Due to their positive features, such as higher efficiency, longer life cycles, and higher energy density, Li-Ion technologies are regarded as the primary form energy storage [1]. It is necessary to understand, analyze, and avoid battery degradation since the batteries' long lifespan is crucial. The degradation of Li-Ion occurs due to irreversible transformations in the composition of cells, such as negative electrode reduction and positive electrode oxidation, brought on by frequent overheating, overcharging, charging, and discharging cycles [2].

Research reveals that compared to the same charging cycle, the discharging procedure typically leads to more damage [3]. The state of health (SOH) parameter is the best tool for gaging battery degradation. SOH refers to battery status measurement on the basis of its designed or nominal characteristics. SOH is measured using three different types of indicators: SOH_p, which is based on peak power, SOH_r, which is based on internal resistance, and SOH_c, which is based on capacity. Among

A. Metya · M. Shadan
KPIT Technologies Ltd., Pune, India

G. Anusha (✉)
KPIT Technologies Ltd., Bangalore, India

A. Haldorai et al. (eds.), *6th EAI International Conference on Big Data Innovation for Sustainable Cognitive Computing*, EAI/Springer Innovations in Communication and Computing, https://doi.org/10.1007/978-3-031-54696-9_3

them, SOH_c can represent the battery's energy storage capacity, much like how the capacity of fuel tank of a conventional fuel vehicle does. SOH_c has been employed in this chapter, and it will be referred to as SOH in subsequent literature.

Recognizing a battery's SOH makes it possible to concentrate on enhancing the overall battery performance and possibly increasing the overall battery life [4]. The optimal battery performance is therefore dependent on SOH prediction in various applications, which require a longer battery lifecycle like electric vehicles (EV) [5]. SOH_c has been employed in this study article, and it will be referred to as SOH in subsequent literature. According to IEEE standard 1188.1996, batteries must be decommissioned whenever their capacity drops to 80%. Hence, the battery's remaining useful life (RUL) is created from SOH (state of health); the cycle in which SOH was downgraded to 80% is known as RUL. SOH/RUL prediction and estimation have been the subject of numerous studies. In the following part, prior research on SOH and RUL prediction is briefly discussed.

2 Literature Review

There is a wealth of literature on the topic of SOH prediction and estimate. In [6], the authors provide a comprehensive analysis of several battery SOH estimating techniques. They classify these methods into three groups: data-driven methods, which are more contemporary, and empirical and model-oriented approaches, which represent more conventional methodologies. In order to accurately assess the deterioration of batteries after several testing methods like impedance, internal resistance, and capacity levels [7–12], empirical methodologies often rely on offline and lab investigations. Both the time-consuming assessment process and the experimental methods' incapacity to provide real-time SOH forecasts are major drawbacks.

Model-based approaches build frameworks, which precisely define the behaviors of battery SOH by using various indications. The methods that are often used are Observers, equivalent circuit models (ECM), and Kalman filters. The twin extended Kalman filter is said to have a problematic battery life, as detailed in reference [13]. This filter is often used for making predictions about SOH and State of Charge (SOC). As discussed in [14], a multi-frame scale observer is fundamental for dual-SOH and an estimation of SOC. Although model-based SOH estimates have a history of producing results with high levels of accuracy and precision, they can also be quite variable and parameter dependent and frequently come at a high computing cost.

The goal of data-driven SOH-estimate approaches, or more precisely machine learning techniques, is to integrate the benefits of the two conventional approaches. They are intrinsically dependent on these conventional approaches since they relied on the data accuracy, which is employed in training the approximating models, which must be gathered by observations or models. Irrespective of this, they offer good prediction precision and straightforward implementation. For SOH prediction, many different machine learning techniques are employed. Artificial Neural

Networks (ANN), Support Vector Machines (SVMs), and Fuzzy Logic [15–17] have all been used in prior research. With various degrees of accuracy, other approaches including random forest and regression tree models are incorporated in [18].

This work presents a precise technique, based on XGBoost, for forecasting the SOH and RUL of a battery. The approach relies on extracting features from the battery charging characteristics. The data used in this research is of the utmost significance in the formulation of concepts and models, as well as in the verification and analysis of outcomes, as is the case with all methodologies based on machine learning. The Stanford Toyota Battery data set is utilized in this research analysis [19].

The below are the key advantages of this approach:

1. By increasing the quantity of batteries used from the whole data set, a bigger and more varied data set will be accessible for validating, training, and evaluating the performance of the model.
2. Using Feature Extraction, to avail use of distinct and identifiable trends in the charging patterns, thereby reduces the quantity of the input data needed.
3. Due to the necessity for reduced model input data, a smaller and less complex XGBoost algorithm is used in this investigation.

3 Problem Statement

The SOH is a widely applied battery degradation indicator, but it tends to lack a widely recognized definition. Some interpretations rely on battery specifications like voltage, self-discharge, and internal resistance. One of the most used definitions relies on the battery capacity, and we define it in the way mentioned below:

$$\mathrm{SOH}(\%) = \frac{C_{\mathrm{i}}}{C_{0}} \times 100 \tag{1}$$

In Eq. (1), C_0 is the cell's initial capacity and C_{i} represents the obtain cycle I capacity. It is known that EOL (End of Life) of the battery occurs when the SOH falls below 80%, at which point performance and consistency are deemed insufficient.

The fundamental idea behind machine learning-based algorithms for predicting the SOH is the fact that it is feasible to analyze variations in observed parameters, such as current and voltage, and accurately deduce battery deterioration using readily available real-time data. More specifically, the Xgboost model is suggested to ascertain the battery SOH for each cycle, which implies that in order to train the network, data on cycle parameters is required. As supervised machine learning is being considered, the ideal battery data set should include the necessary capacity or SOH values for training the prediction model.

As a result, Stanford battery data is being used for this analysis. 47 Li-Ion batteries with 1.1 Ah, an estimated capacity, are used to gather information. Capacity measurements and charging and discharging procedures are part of every cycle. The CC-CV (constant current-constant voltage) is employed by the fixed charging approach. This entails charging the battery at varying currents in many stages until the time when the voltage of the battery is 3.7 V. Currently, the voltage is maintained at a constant level until the cell current falls below 20 mA. Every cycle's data set includes the battery capacity evaluation that was taken throughout the discharge process. All of the cells in the data set have their degradation per cycle shown here.

3.1 Feature Extraction

The degradation curves for the battery capacity resemble the optimal results for proposed prediction model. As input to the model, seven features were obtained from the phase of charging the battery. The primary benefit of prioritizing the method of charging is that it leverages a universal charging protocol, enabling efficient identification, retrieval, and deployment of qualities included in the data of all battery cells. Since various discharge circumstances impact the evaluation of the battery characteristics during charging, it is vital to mention that these impacts are automatically incorporated. As a result, the proposed traits are used to evaluate the degradation of batteries. Because all the curves are so similar in shape, this method produces duplicate data yet is more accurate and easier to implement than sampling. This chosen attribute is shared by the current and voltage curves because it shows when the CC-CV recharging process goes from an even current to a steady voltage condition [20]. There are vital differences in V-I charging curves between successive cycles. The I-C curve, which represents incremental capacity, also captures the other features.

3.2 Calculation of IC-Curve

The incremental capacity (IC) curve is derived by deducting battery charge variations from those in the voltage at battery's terminals. The following equation is used to compute the IC curve:

$$\frac{dQ}{dV} = \frac{Q_k - Q_{k-N}}{V_k - V_{k-N}} = \frac{\int_{t_{k-N}}^{k} i\,dt}{V_k - V_{k-N}} \tag{2}$$

where t_k, V_k, and Q_k represent the time at kth point of sampling, the voltage at the battery terminal, and the stored battery charge, respectively. The differential window is denoted by N. The charging current is denoted by the symbol i. When the IC curve is computed using the constant current (CC) charging method and the sampling frequency remains constant, the calculation below may be abbreviated as follows:

$$\frac{dQ}{dV} = \frac{\int_{t_{k-N}}^{k} i dt}{V_k - V_{k-N}} = \frac{I(t_k - t_k - N)}{V_k - V_{k-N}} = I.\Delta t.\frac{N}{V_k - V_{k-N}} \tag{3}$$

when I stand for the procedure of charging with a steady current. This is the sample interval, denoted by t. We have two constants, Δt and I. During CC charging, the equation states that the IC curve's slope is only relevant to the presumed permission of the slope of the charging voltage curve. There will be a peak on the IC curve for slowly changing voltages. An IC curve crest indicates a rapid increase in voltage. Alterations to SOH have a direct bearing on the features of voltage changes. This means that the IC curve, rather than the voltage curve, represents the characteristics of voltage variations.

Throughout the CC charging period, the rate of voltage growth slows down at a key point when the SOH of a Li-Ion battery decreases. At this stage, the CC charging cycle is completed sooner since the voltage is greater. As the battery's SOH declines, the voltage will climb quicker while charging continuously using the same power source. Conversely, a battery's ability to store energy declines with time when its SOH decreases. The slower the increase in voltage, the larger the value of $\frac{dQ}{dX}$. It has been shown that larger batteries have a reduced voltage at the beginning of the voltage level, when the voltage rises slowly. Also, the battery's SOH is worse when the area between the IC curve and abscissa is smaller during the voltage range of 2.7–3.7 V. As a result, the IC curve has SOH aging data that can be utilized to predict the SOH. As a result, six features from IC curve are computed such as IC curve Peak, IC curve Peak Voltage, Area under the IC curve at voltage bins from 3.1 to 3.2 V, 3.2 to 3.3 V, 3.3 to 3.4 V, and 3.4 to 3.5 V.

4 Methodology

The prediction method proposed in this study is effectively described by the XGBoost algorithm. The workflow of the RUL forecasting methods for Li-Ion battery is shown in Fig. 1 to prove it further clearly. Data processing, model testing, and training are the three procedures in this analysis.

Data processing is done in Phase I, where it includes data simulation, preprocessing, and data aggregation needed for this analysis. In this process, raw data was analyzed from different open sources and a few more columns will be appended to data by using some functions in preprocessing. The attributes are taken from the preprocessed data for the data aggregation on all the batteries. The aggregated data will be used then for the model training analysis.

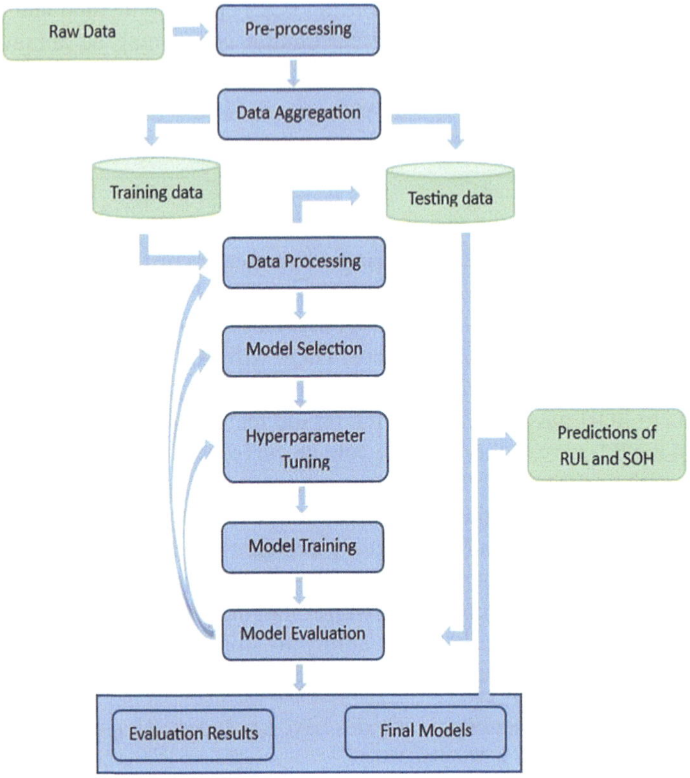

Fig. 1 Methodology or workflow for this analysis

Phase II is training the model, which entails selecting the model by determining its hyperparameters using a cross-validation research. Testing, in this Phase III, predictions of RUL and SOH are evaluated by using the models created from model training phase and these were evaluated by performance metrics.

4.1 Data Simulation

The data set was collected and analyzed from three different open sources.

1. Stanford Toyota Battery Data
2. NASA PCOE (Prognostics Center of Excellence) Battery Data
3. NASA Random walk Battery Data

4.2 NASA PCOE (Prognostic Center of Excellence) Battery Data

This repository's data set includes information from thirty-four 18,650 lithium-ion batteries. Throughout the cycling procedure, these batteries, which have a nominative capacity of 2.0 Ah, were exposed to various discharge levels and environmental temperature circumstances. Each cycle involves a process of charging and discharging, together with the collection of impedance measurements using EIS. To charge the battery using the fixed charging method, the current must remain constant at 1.5 A until the voltage of the cell reaches 4.2 V. When the cell current falls below 20 mA, the charging mechanism shifts to keeping the voltage constant. The protocol used for this procedure is often referred to as the CC-CV protocol.

The main difference between a charging and discharging situation is the cut-off voltage and the variable current. Based on these features, the collection may be sorted into many groups. There are different batteries in separate groups. These groups are formed based on varying discharge currents. For example, group 1 is discharged at 2 A, whereas group 2 is discharged at 4 A, and so on. These groups are subjected to a constant current (CC) discharge at their corresponding currents. The following Table 1 provides samples of various working situations for a selection of batteries, together with the corresponding number of possible cycles. Even though group 1 offers a lot more cycles, group 2 undergoes a more severe discharge process that uses more temperature and current.

4.3 Stanford Toyota Data

A total of 124 batteries with lithium ion that were tested under fast-charging parameters until they failed make up the data set. This data is available to download as JSON and MAT files. This data set will be consisting of two diverse types such as cycle level data and summary level data, where cycle level data stands for whole information, while summary level data stands for the summary of the data by

Table 1 Battery operating conditions

				Discharge	
Group	Battery	Temp	Cycles	Current	Cut-off voltage
1	B0005	24	168	2	2.7
	B0006	24	168	2	2.5
	B0007	24	168	2	2.2
	B0018	24	132	2	2.5
2	B0029	43	40	4	2.0
	B0030	43	40	4	2.2
	B0031	43	40	4	2.5
	B0032	43	40	4	2.7

Table 2 Sample row data

Cycle index	Step type	Charge capacity	Discharge capacity
1	Discharge	1.0644852	0.992948323

Table 3 Sample row data

Current	Voltage	Temperature	Internal resistance
−4.40001	2.8	36.82826233	0.018386904

performing group by on cycle index column. Cycle level data, i.e., raw data, consists of eight columns such as cycle index, step type, charge capacity, discharge capacity, current, voltage, temperature, and internal resistance. The format of data having sample rows is shown below.

Tables 2 and 3 represent the sample raw data that is being used. Cycle index, in Tables 2 and 3, represents consecutive discharge and charge cycle in order by same cycle number, step type indicates whether the battery is in charging state or discharging state where current will be negative when discharge is happening and current is positive when charge is happening, *charge capacity remains constant throughout the cycle when discharge is happening* i.e., step type is discharge.

Figure 2 represents the two different batteries (el150800460511_CH37 and el150800460513_CH41), during different cycles where *x-axis* is *cycle index* and *y-axis* are *voltage and current.*

4.4 Data Processing

Further columns, including the relative time and absolute time, are added to the raw data from the analysis done above, where absolute time is generated by using the summary level data which has charge duration of each cycle for every battery and relative time is the time in seconds. The sample data of these two other features is shown in Table 4.

For the analysis, cell level data was considered from Stanford Toyota. Some interpretations are done on data by using exploratory data analysis.

The data set used for this research comprises 124 individual cell data points, including a wide variety of cycles and involving 72 unique fast-charging settings. The lifespan of a cycle or end of life is determined by the total number of cycles until the cell reaches 80% of its intended capacity. This Li-Ion rechargeable cell is capable of high power efficiency and cycling in temperature-controlled environment, which has nominal capacity of 1.1 Ah under different fast-charging conditions (~10 min charging).

Figure 3 stands for the capacity plot for all the batteries. Based on Fig. 3, for further analysis, batteries were considered within *400–600* range of cycle number with few more columns appended to data. Cells are arranged in Parallel and Series to establish a battery pack of configuration *100S20P*.

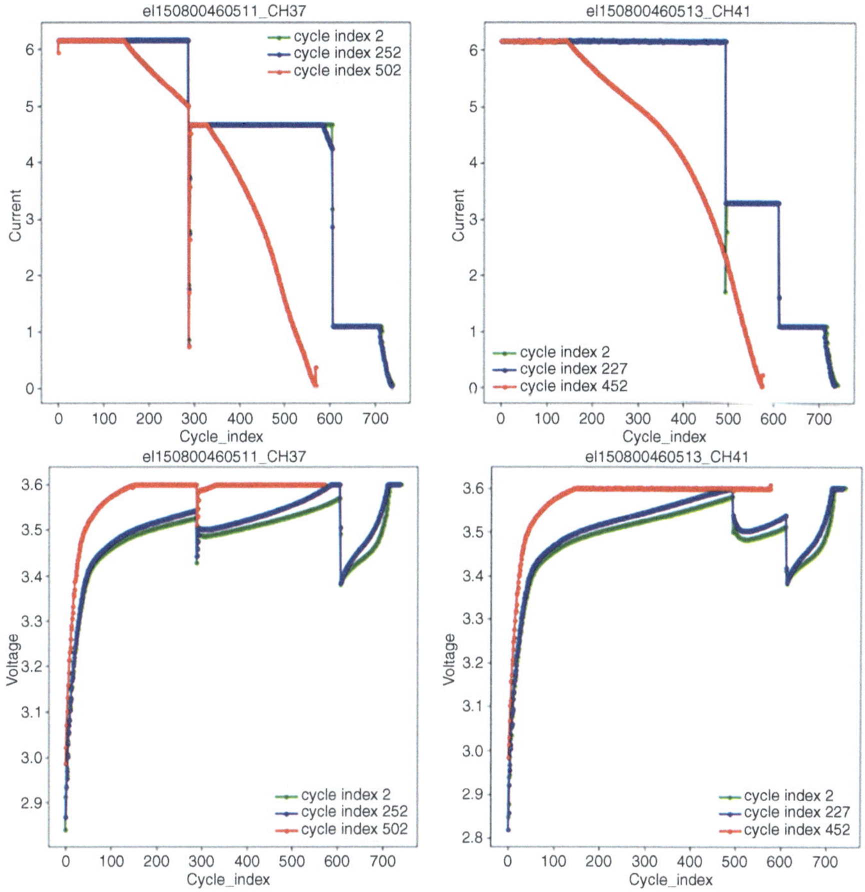

Fig. 2 Current and voltage analysis for batteries

Table 4 Features added to data

Relative time	Absolute time
0	2017-07-02 01:18:56+00:00

Data now includes more attributes, which are generated for every battery. The data with all the required features is shown in Table 5. After getting this data, data will be aggregated for all the batteries.

In Table 5, the parameters that are created for further analysis are listed.

4.5 *Data Aggregation*

Data aggregation is done on all the batteries to get the required data for prediction analysis. The sample data smoothened on one battery is shown in Table 6.

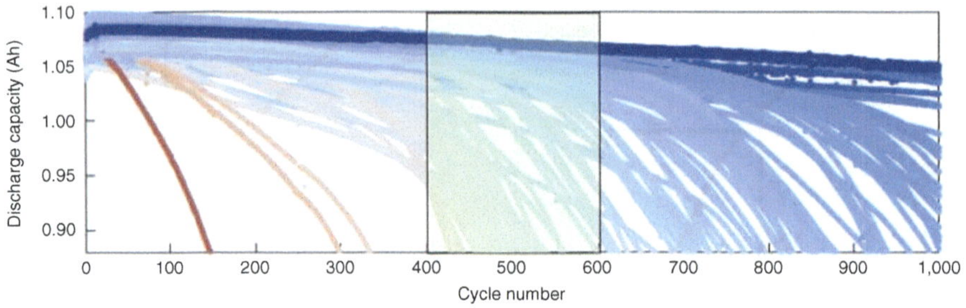

Fig. 3 Capacity plot for all batteries. (Taken from Severson et al. [19])

Table 5 Features needed for all the batteries

Vin	4Y1SL65848Z411482
Battery pack id	El150800460451
Cycle number	1
Step type	Charge
Timestamp actual	2017-07-02 01:18:56+00:00
Time	0.64
Capacity	1.064485
Charge capacity	0.002876
Discharge capacity	3.55E-12
Current	79.19899
Voltage	282.4921
Temperature	30.88699
Soc	0.117186
Odometer	223.4652
Min cell voltage	2.421012
Max cell voltage	2.924921
Avg cell voltage	2.797966
Total charge energy	0.003877
Total discharge energy	0
Min voltage cell id	87
Max voltage cell id	79
Timestamp	1/18/2015 8:38:24 AM
Usage pattern	Infrequent usage

Table 6 represents the smoothened data for one battery after aggregation and the parameters are described below:

Voltage: It represents at what voltage the highest peak of dQ/dV is reached.
Peak: Absolute value of peak dQ/dV.
Seg_3_2_3_3: Area Under Curve (AUC) within voltage range of 3.2–3.3 V.
Seg_3_3_3_4: AUC within voltage range of 3.3–3.4 V.
Seg_3_4_3_5: AUC within voltage range of 3.4–3.5 V.

Table 6 Data with aggregated features

Voltage	Peak	Seg_3_2_3_3	Seg_3_3_3_4	Rated Wh
3.435	110.9	0.0037785	1.57660458	3.85
cc-Time	cc-Current	Seg_3_4_3_5	Seg_3_5_3_6	Rated Ah
188.75	8.799	10.7734484	9.142986873	1.1

Seg_3_5_3_6: AUC within voltage range of 3.5–3.6 V.
cc-Time: This is the longest constant current (CC) time to be considered.
cc-Current: This is constant charging (CC), absolute value of constant current (A).

4.6 Method of Data Aggregation

Data aggregation is a crucial part. First, time series data for each individual cycle is analyzed. However, in this analysis, the developed method is designed in a way such that single aggregated feature is required for each individual cycle for that model. The physical importance of these aggregated features has already been mentioned in the preceding sections. In this section, the methodology for extracting these aggregated features for individual cycles is discussed.

4.7 CC (Constant Current)

There are two aggregated features concerning the battery's constant current phase. The RLE (Run Length Encoder) method is employed to determine the CC-Phase (Constant Current) from a time series. The Stanford batteries are charged in multiple constant current phases, as shown in Fig. 4. As a result, the longest CC Phase is identified by utilizing this RLE Method.

In Table 6, the current of that CC-Phase and the duration of that CC-Phase are identified as two prominent features and are denoted as cc-current and cc-time, respectively.

4.8 Features of IC-Curve

IC-Curve was used to generate a total of six features. The physical relevance of the IC curve has already been discussed. Individual cycles current and voltage time series are used to generate IC-Curve. To obtain the $\frac{dQ}{dV}$ values, apply Eq. (2) or Eq. (3) to the current and voltage time series. A customized function ic_curve.py is used for $\frac{dQ}{dV}$. This function takes three arguments: start voltage, end voltage, and voltage

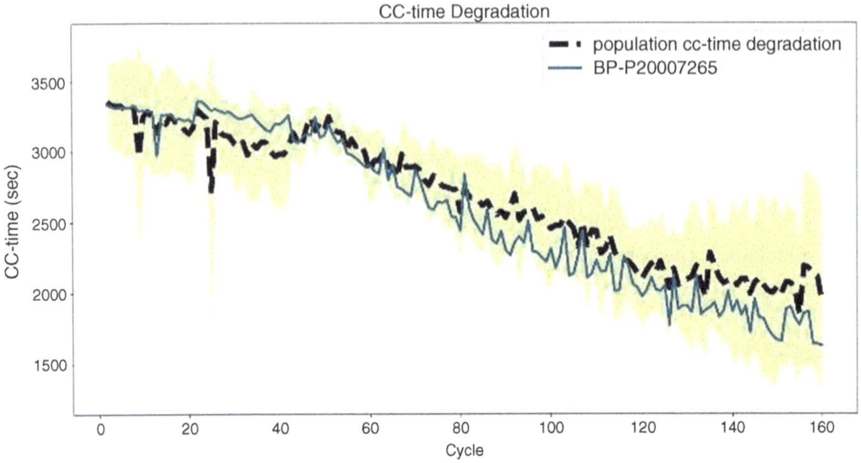

Fig. 4 CC-time degradation analysis

bin where start voltage and end voltage represent CC-Phase start and end voltages. The default voltage bin value is 0.005 V. These values can be changed as needed by the user. These $\frac{dQ}{dV}$ values are then smoothed using the Gaussian filter. The smoothed $\frac{dQ}{dV}$ is now plotted against the corresponding voltages that are used as two features and are referred to a voltage and peak.

The graph below depicts the evolution of peak and voltage characteristics as the cycle number increases. The voltage feature value gradually rises as the battery degrades and rapidly rises as the battery ages. Peak feature value drops dramatically as the battery ages (Figs. 5, 6, and 7).

Also, the area under curve (AUC) at four different voltage bins, i.e., 3.2–3.3 V, 3.3–3.4 V, 3.4–3.5 V, and 3.5–3.6 V, are calculated. The change in pattern of these AUCs with different cycle numbers are shown in Fig. 8.

Based on these aggregated data generated, few plots were analyzed such as AUC and capacity degradation plots.

Figure 8 is the area under curve for few batteries, where *x-axis* is the *cycle number* and *y-axis* is the *area under curve (AUC)* for battery.

Figure 9 is the degradation plot for capacity where *x-axis* is the *cycle number* and *y-axis* is *capacity* in Ah.

Figure 10 represents the degradation plot for constant current (CC) over cc-time, where *x-axis* is the *cycle number* and *y-axis* is the *cc-time*.

5 Model Training

XGBoost is an effective approach utilized for both regression and classification tasks. XGBoost utilizes the gradient-boosting framework to iteratively add decision trees and enhance the learning process by fitting values with numerous residual

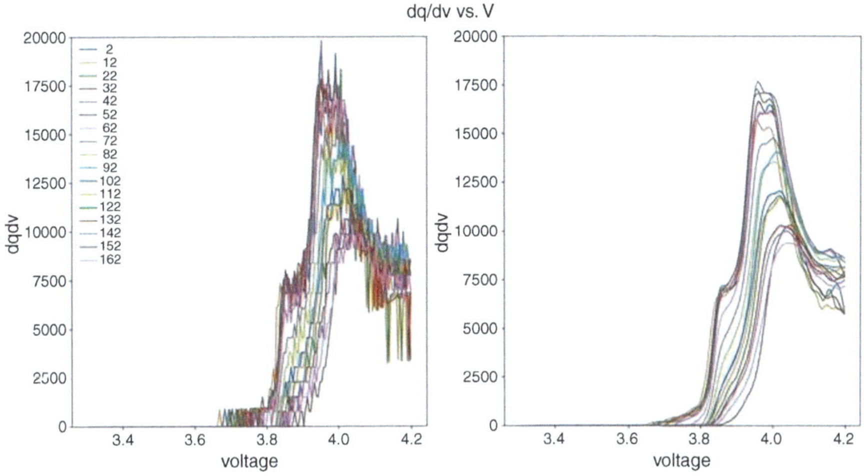

Fig. 5 *dQ/dV* and voltage plot

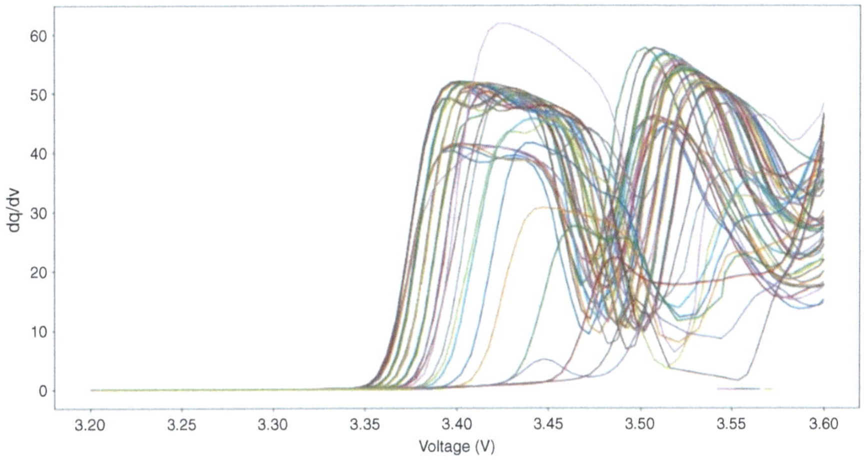

Fig. 6 *dQ/dV* and voltage plot for Stanford Toyota Data

rounds. Unlike gradient boosting, XGBoost employs a Taylor expansion to compare the gradient descent. In order to achieve more accuracy, the method often utilizes a reduced number of decision trees because of its improved balance between variance and bias. To accurately predict a battery's RUL, XGBoost is used to build a linkage between the extracted attributes.

The model trained on multiple batteries which correctly forecasted the RUL and SOH. The test data for the battery sets el1508004605608CH31, el15080046035CH13, el150800460501CH22, el15800460526CH40, and el150800460511CH37 are represented in this analysis. The Li-Ion battery capacity barrier for this analysis was set at 1.1 Ah. Battery life ends when the performance reaches a specified failure value (End of Life (EOL)). The usual failure level is 80% of the rated capacity. The XGBoost

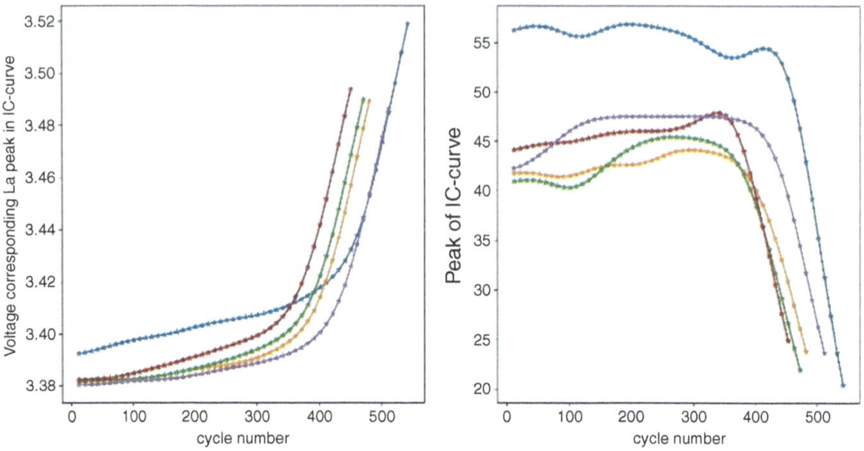

Fig. 7 IC-Curve peak and voltage corresponding to IC-Curve Peak

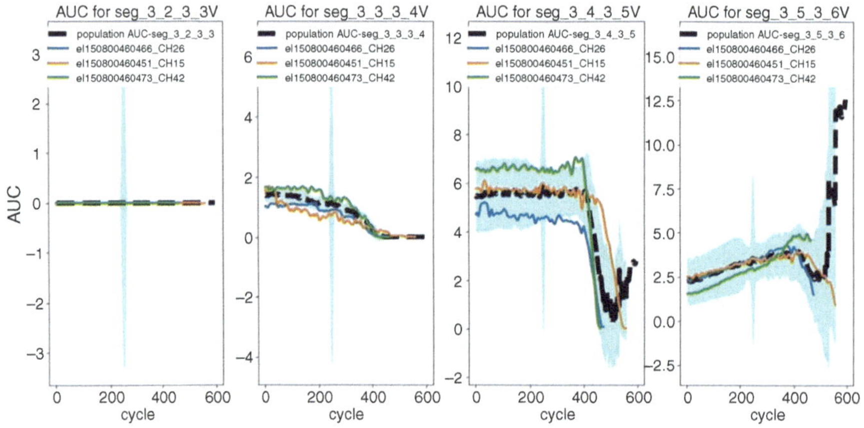

Fig. 8 AUC plots at different voltage ranges

model was used as the prediction algorithm in this case. The characteristics collected from the prior approach were employed as inputs meant for model training.

Parameters used for XGBoost Model are described in Table 7.

In Table 7, the hyperparameters used in model training are listed.

5.1 Results

After the analysis of model training, best parameters of XGBoost Model were identified which are shown below.

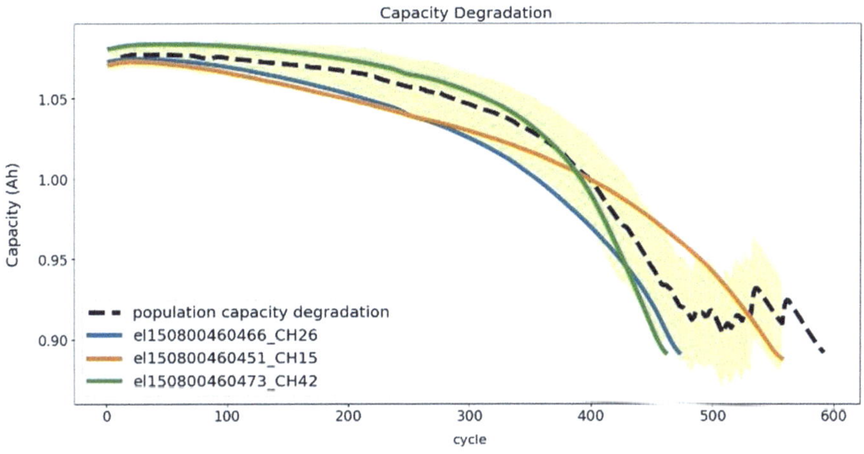

Fig. 9 Capacity degradation plot

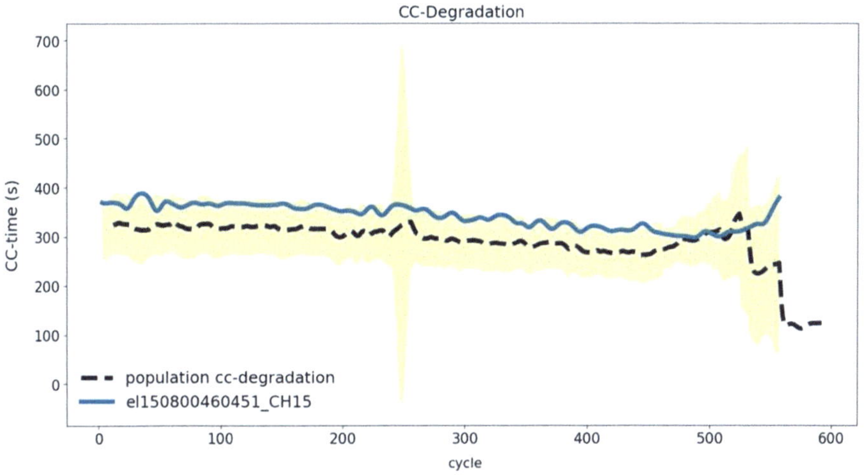

Fig. 10 CC-degradation plot

```
XGBRegressor (base_score =0.5, booster='gbtree',
             callbacks=None, colsample_bylevel=1,
             colsample_bynode=1, colsample_bytree=1,
             early_stopping_rounds=None,
             enable_categorical=False,
             eval_metric=None, gamma=0,
             gpu_id=-1, grow_policy='depthwise',
             importance_type=None,
             interaction_constraints ='',
             learning_rate=0.300000012,
```

Table 7 XGBoost model parameters

Parameter	XGBoost values
n_estimators	[100, 500, 900, 1100, 1500]
learning_rate	[0.05, 0.1, 0.15, 0.20]
max_depth	[2, 3, 5, 10, 15]
booster	gbtree, gblinear
base_score	[0.25, 0.5, 0.75, 1]
min_child_weight	[1, 2, 3, 4]

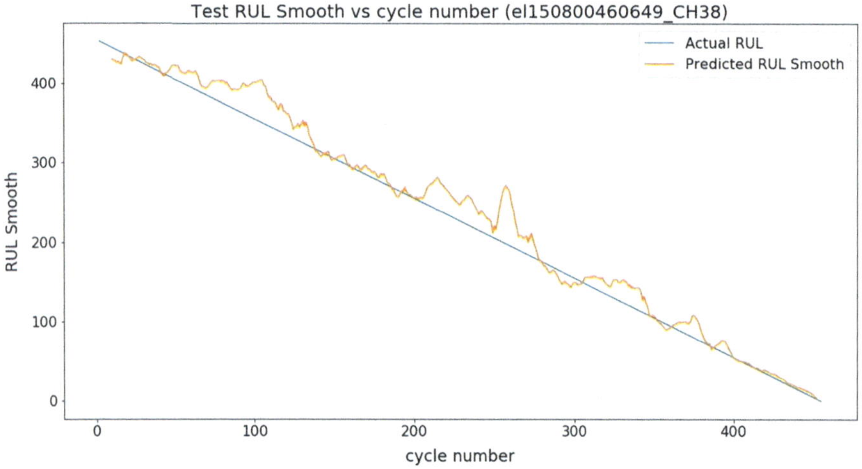

Fig. 11 Predicted RUL and actual RUL (el150800460649_CH38)

```
max_bin=256, max_cat_to_onehot=4,
max_delta_step=0, max_depth=6, max_leaves=0,
min_child_weight=-1, missing=nan,
monotone_constraints=" ()',
n_estimators=100, n_jobs=0, num_parallel_tree=1,
objective='reg: squarederror',
predictor='auto,' random_state=0, reg_alpha=0, ...))]
```

The above hyperparameters of XGBoost Model are considered as best parameters.

Figure 11 illustrates the influence of the recommended approach on the battery RUL on the basis of goof prediction where *x-axis* is *cycle number* and *y-axis* is *RUL* (remaining useful life).

Figure 12 represents the graph for actual and predicted data where *x-axis* is the *cycle number* and *y-axis* is the *SOH* (state of health).

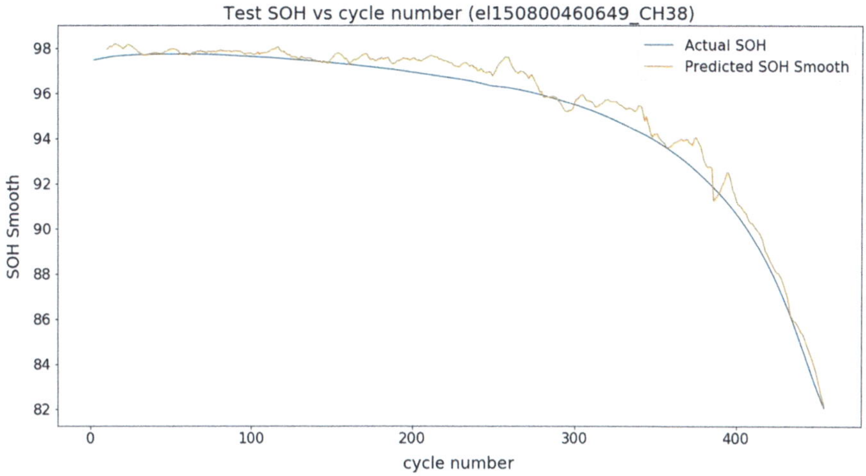

Fig. 12 Actual SOH and predicted SOH (el150800460649_CH38)

5.2 *Evaluation Metric*

The Root Mean Square Error (RMSE) and Mean Absolute Error (MAE) were used to gauge how accurate the suggested method's forecasts were.

$$\mathrm{RUL}_{\mathrm{RMSE}} = \sqrt{\frac{1}{N}\sum_{n=1}^{N}\left(\mathrm{RUL}_{n} - \mathrm{RUL}_{\hat{n}}\right)^{2}} \tag{4}$$

$$\mathrm{RUL}_{\mathrm{MAE}} = \sqrt{\frac{1}{N}\sum_{n=1}^{N}\left|\mathrm{RUL}_{n} - \mathrm{RUL}_{\hat{n}}\right|} \tag{5}$$

$$\mathrm{SOH}_{\mathrm{RMSE}} = \sqrt{\frac{1}{N}\sum_{n=1}^{N}\left(\mathrm{SOH}_{n} - \mathrm{SOH}_{\hat{n}}\right)^{2}} \tag{6}$$

$$\mathrm{SOH}_{\mathrm{MAE}} = \sqrt{\frac{1}{N}\sum_{n=1}^{N}\left|\mathrm{SOH}_{n} - \mathrm{SOH}_{\hat{n}}\right|} \tag{7}$$

$$R^{2} = 1 - \frac{\Sigma\left(y_{i} - y_{\hat{i}}\right)^{2}}{\Sigma\left(y_{i} - \overline{y}_{i}\right)^{2}} \tag{8}$$

In Table 8, the results from the training model XGBoost for 15 batteries are given. (*Train*: el150800440551_CH18, el150800460449_CH29, el150800460451_CH15, el150800460466_CH26, el150800460473_CH42, el150800460474_CH44, el150800460478_CH16, el150800460480_CH17, el150800460492_CH34,

Table 8 Results from XGBoost model for 15 batteries

Results	RMSE	MAE	R^2
RUL	44.72713	32.58512	0.891399
SOH	1.361089	0.881814	0.909074

Table 9 Results from XGBoost model for 36 batteries

Results for RUL training data	
RMSE	10.822074
MAE	8.8269671
R^2	0.9938214
Results for RUL testing data	
RMSE	49.684139
MAE	32.051394
R^2	0.8685912
Results for SOH training data	
RMSE	0.2270661
MAE	0.1575968
R^2	0.9973853
Results for SOH testing data	
RMSE	1.3640772
MAE	1.0226814
R	0.9149614

el150800460498_CH43, *Test*: el150800460501_CH22, el150800460511_CH37, el150800460510_CH27, el150800460513_CH41, el150800460526-CH40).

In Table 9, the results for all the 36 batteries with training and testing accuracies for RUL and SOH are given.

6 Conclusion

This research proposes a fast and precise real-time prediction method that is effective for electric vehicles (EVs) real-time operation. Differentiating the voltage curves of the batteries when they are charged under CC (constant current) circumstances is essential for obtaining the IC (incremental capacity) curve. A Gaussian filter is employed to remove the noise in the IC Curve. As crucial features, the IC-Curve's peak voltage, peak value, and area under curve at a few distinct voltage bins are all taken into consideration. Lastly, the battery data from the Stanford University aging experiment is employed to validate the reliability of the approach suggested in this chapter. The empirical findings prove that the RUL or SOH prediction MAE (Mean Absolute Error) suggested in this chapter is low with extremely high accuracy ($R^2 > 0.91$).

The distinctiveness and originality of this study reside in the following aspects:

1. To achieve a prominent level of accuracy in the prediction of SOH, this work creatively combines IC Curve with XGBoost.
2. The research makes use of a technique that, using only data from a steady-state charging condition, can reliably predict the SOH. When compared to another method that uses a combination of charging and discharge data to estimate SOH, this one requires less processing power.
3. The approach employed in this analysis of SOH estimation is created in a way that it can be deployed in "clouds" and used in real-time. This improves this method's scalability, usefulness, and practicability.

7 Future Scope

Even though the techniques suggested in this chapter has produced accurate SOH and RUL predictions, few aspects still require improvements.

1. The training phase consumes a lot of energy since it requires data from the whole battery life cycle for training.
2. Due to the lack of the battery aging data at various temperatures, it is impossible to describe how the temperature may affect the SOH or RUL prediction approach used in this analysis.
3. The data sets used in this analysis were produced through experiments in labs under controlled conditions, which may differ from the actual real-life data sets. Hence, training using real-world data sets is crucial.

References

1. Ghassan Zubi, Rodolfo Dufo-Lopez, Monica Carvalho, Guzay Pasaoglu, "The Lithium-Ion Battery: State of the Art and Future Perspectives," Renew. Sustain. Energy Rev. (ISSN: 1364-0321) 89 (2018) 292–208.
2. Nikita S. Popov, Vladislav I. Anibroev, Mikhail M. Mosin, "Study of Processes that cause degradation of Lithium-Ion Batteries," In: 2021, 3rd International Youth Conference on Radio Electronics, Electrical and Power Engineering, Reepe, 2021, PP. 1–4.
3. Feng Lin, Dennis Nordlund, Tsu-Chein Weng, Ye Zhu, Chunmei Ban, Ryan M. Richards, Huolin L. Xin, "Phase Evolution for Conversion Reaction Electrodes in Lithium-Ion Batteries," Nature Commun. 5(1) (2014).
4. Huixin Tian, Pengliang Qion, Kun Li, Zhen Zhao, "A Review of the State of Health for Lithium-Ion Batteries: Research Status and Suggestions," J. Cleaner Prod. (ISSN: 0959-6526) 261 (2020) 120813.
5. M.S. Hossain Lipu, M.A. Hannan, Aini Hussian, M.M. Hoque, Pin J. Ker, M.H.M Saad, Afida Ayob, "A Review of State of Health and Remaining Useful life Estimation Methods for Lithium-Ion Battery in Electric Vehicles: Challenges and Recommendations," J. Cleaner Prod. (ISSN: 0959-6526) 205 (2018) 115–133.
6. Nassim Noura, Loic Boulon, Samir Jemei, "A Review of Battery State of Health Estimation Methods: Hybrid Electric Vehicle Challenegs," World Electr. Veh. J. 11 (4) (2020) 66.

7. Hans-Georg Schweiger, Ossama Obeidi, Oliver Komesker, Andre Raschke, Michael Schiemann, Christian Zehner, Markus Gehnen, Michael Keller, Peter Brike, "Comparison of several methods for determining the internal resistance of Lithium-Ion cells," Sensors (ISSN: 1424-8220) 10 (6) (2010) 5604–5625.
8. Grzegorz Pilatowicz, Andrea Marongiu, Julia Drillkens, Philipp Sinhuber, Dirk Uwe Sauer, "A critical overview of definitions and determination techniques of the internal resistance using Lithium-Ion, Lead-acid, Nickel metal-hydride batteries and electrochemical double-layer capacitors as examples," J. Power Sources 296 (2015) 365–376.
9. Matteo Galeotti, Lucio Cina, Carrado Giammanco, Stefano Cordiner, Aldo Di Carlo, "Performance Analysis and SOH (State of Health) Evaluation of Lithium Polymer Batteries through Electrochemical Impedance Spectroscopy," Energy (ISSN: 0360-5442) 89 (2015) 678–686.
10. Yingzhi Cui, Pengjian Zuo, Chunyu Du, Yunzhi Gao, Jie Yang, Xinqun Cheng, Yulin Me, Geping Yin, "State of Health Diagnosis Model for Lithium-Ion Batteries Based on Real-Time Impedance and Open Circuit Voltage Parameters Identification Method," Energy. (ISSN: 0360-5442) 144 (2018) 647–656.
11. Xiaoyu Li, Zhenpo Wang, Lei Zhang, Chanfu Zou, David. D. Dorrell, "State of Health Estimation for Li-Ion Batteries by Combining the Incremental Capacity Analysis Method with grey relational analysis," J. Power Sources (ISSN: 0378-7753) 410–411 (2019) 106–114.
12. Xiaoyu Li, Changgui Yuan, Xiaohui Li, Zhenpo Wang, "State of Health Estimation for Li-Ion Battery Using Incremental Capacity Analysis and Gaussian Process Regression," Energy (ISSN: 0360-5442) 190 (2020) 116467.
13. Nikolas Wassiliadis, Jorn Adermann, Alexander Frericks, Mikhail Pak, Christoph Reiter, Boris Lohmann, Markus Lienkamp, "Revisiting the Dual Extended Kalman Filter for Battery State of Charge and State of Health Estimation: A Use-Case Life Cycle Analysis," J. Energy Storage (ISSN: 2352-152x) 19 (2018) 73–87.
14. Changfu Zou, Chris Manzie, Dragan Nesic, Abhijit G. Kallapur, "Multi-Time-Scale Observer Design for State of Charge and State of Health of a Lithium-Ion Battery," J. Power Sources (ISSN: 0378-7753) 335 (2016) 121–130.
15. Verena Klass, Marten Behm, Goran Lindbergh, "A Support Vector Machine Based State of Health Estimation Method for Lithium-Ion Batteries under Electric Vehicle Operation," J. Power Sources (ISSN: 0378-7753) 270 (2014) s 262–272.
16. Jinhao Meng, Lei Cai, Guangzhao Luo, Daniel-loan Store, Remus Teodorescu, "Lithium-Ion Battery State of Health Estimation with Short-term Current Pulse Test and Support Vector Machine," Microelectron. Reliab. (ISSN:0026-2714) 88–90 (2018) 1216–1220.
17. Xin Sui, Shan He, Daniel-loan Stroe, Remus Teodorescu, "State of Health Estimation for Lithium-Ion Battery Using Fuzzy Entropy and Support Vector Machine," 2020 IEEE 9th International Power Electronics and Motion Control Conference, IPEMC2020-ECCE Asia, PP. 1417–1422.
18. Huaging Xu, Yanqing Peng, Lumei Su, "Health State Estimation Method of Lithium Ion Battery Based on NASA Experimental Data Set" In: IOP Conference Series: Material Science and Engineering, 452, IOP Publishing, 2018, 032067.
19. Kristen A. Severson, Peter M. Attia, Norman Jin, Nicholas Perkins, Benben Jiang, Zi Yang, Michael H. Chen, Muratahan Aykol, Patrick K. Herring, Dimitrios Fraggedakis, Martin Z. Bazant, Stephen J. Harris, William C. Chueh and Richard D. Braatz, "Data-Driven Prediction of Battery Cycle Life Before Capacity Degradation," Nature Energy, 2019.
20. Driscoll, L. de la Torre, S. and Gomez-Ruiz, J.A., 2022. "Feature-Based Lithium-Ion Battery State of Health Estimation with Artificial Neural Networks," Journal of Energy Storage, 50, P. 104584.

Integrating Fuzzy Logic and Deep Learning for Effective Network Attack Detection with Fuzzified Deep Convolutional Neural Network

D. Venkatesh, T. Saravanan, and S. Vasundra

1 Introduction

An intrusion detection system (IDS) keeps track of unusual activity and distinguishes among regular and illegal (intrusive) behavior in a host system or network. Intrusion prevention systems defend assets and information against unauthorized entry, abuse, and hostile incursions in distributed computer environments. NN, SVM, NB, and other machine learning techniques are used for identifying intrusions. The entire system is continuously monitored and examined by IDSs, which enables the ML to identify typical/everyday activity [1]. This enables the model to recognize unusual or aberrant behavior and respond accordingly. IDS testing and development utilize the KDD99 dataset [2]. The IDS is responsible with keeping an eye on and analyzing network operation in order to distinguish among typical and abnormal actions [3]. The reliability and dependability of a computing platform might possibly suffer serious harm if the unusual behavior is left undiscovered. Therefore, it is essential to increase the rate of aberrant activity recognition.

In addition to detecting abnormal behavior, the IDS also has to reduce the total amount of false positives to prevent unnecessary effort and misunderstanding. False positives don't endanger the system [4]. The IDS's capacity for offering accurate and dependable findings may be compromised, however, if they happen often. The key to making an IDS function properly is to find a good mix among the number of things it finds and the number of false positives it gives [5]. As regular and unusual

D. Venkatesh (✉) · T. Saravanan
GITAM School of Technology, GITAM (Deemed to be University), Bengaluru, India

S. Vasundra
Jawaharlal Nehru Technological University, Anantapur, India
e-mail: vasundras.cse@jntua.ac.in

A. Haldorai et al. (eds.), *6th EAI International Conference on Big Data Innovation for Sustainable Cognitive Computing*, EAI/Springer Innovations in Communication and Computing, https://doi.org/10.1007/978-3-031-54696-9_4

activities shift, integrating recognition with rates of false positives is harder. The network's activity may vary, and the intrusion detection system needs to be cognizant of such variations and adjust as necessary. If not, the IDS's capacity to provide reliable and exact findings is significantly reduced [6]. As a result, an IDS needs to adjust to various contexts that may introduce various activities and behaviors that the IDS is not capable of detecting. The key contribution of the proposed work is

- Fuzzification of Input Features: The FDCNN incorporates fuzzification techniques to represent input features using linguistic terms, allowing for handling uncertain and imprecise information associated with network traffic data. This fuzzified representation captures the inherent complexity and variability of network attacks, leading to improved detection performance.
- Hierarchical and Abstract Representation Learning: The FDCNN employs a deep convolutional neural network architecture to learn hierarchical and abstract representations of network traffic patterns. By automatically extracting discriminative features from raw network data, the FDCNN enables more accurate and reliable detection of network attacks compared to traditional machine learning approaches.

1.1 Contribution

The significant contribution of this chapter is summarized below:

- The FDCNN method leverages the power of deep learning and fuzzy logic to enhance the accuracy and robustness of network attack detection systems. The FDCNN model integrates fuzzification techniques with a deep convolutional neural network architecture.
- Fuzzification enables the representation of input features with linguistic terms, allowing the network to handle uncertain and imprecise information associated with network traffic data.
- This fuzzified representation helps capture the inherent complexity and variability of network attacks. The deep convolutional neural network component of the model is designed to learn hierarchical and abstract representations of network traffic patterns.
- The network architecture comprises multiple convolutional and pooling layers, followed by fully connected layers for classification. This enables the FDCNN model to automatically learn discriminative features from raw network data, facilitating accurate detection of network attacks.

1.2 *Organization*

The rest of the chapter is organized as follows: Sect. 2 explains the literature review. Section 3 formulates the proposed fuzzified deep convolutional neural network. Section 4 illustrates the experimentation analysis, and result and discussion are also presented in this section. Finally, Sect. 5 concludes the chapter.

2 Literature Review

As demand for Internet connections grows, we must protect them from intruders and attackers. On this part, we have examined different author's surveys with respect to effective and attack-free networks. The author evaluation is presented here with the pertinent information included in the reference's parts.

Information packets in Internet traffic that don't match the established usual profile are identified and analyzed by anomaly-based IDS. Ripon techniques for ML were used by [7] to efficiently identify invasions. ML is an approach to statistics for addressing issues related to regression and classification. These techniques include NB organization, SVMs for analysis and regression, and KNNs for both [8]. In recent years, NIDS research has focused on using ML and shallow learning methods like NBs and DTs. A unique algorithm for deep learning was put out by [9] to allow NIDS functioning inside contemporary networks. The model they suggest combines deep neural networks with shallow learning, thereby making it able to accurately evaluate a variety of Internet data. They especially integrate the speed and effectiveness of RFs with the ability of layering their suggested non-symmetric deep auto-encoder (NDAE). Different authors have successfully investigated their hypotheses using GPU-enabled TensorFlow and their analysis of the KDD Cup '99 and NSL-KDD databases yielded positive outcomes [10].

These sub-DBN algorithms can classify effectively, acquire and explore high-level abstraction features, and dynamically decrease information dimensionality. The fuzzy participation values of each test example in each sub-DBNs classification are determined utilizing the KNN criteria. Based on fuzzy membership weights, the output of each sub-DBNs classifier is combined. On the NSL-KDD and UNSW-NB15 datasets, experimental findings demonstrate that our proposed model outperforms other well-known classification techniques in terms of overall accuracy, recall, precision, and F1-score. In comparison to cutting-edge security breach detection techniques, the suggested model performs superior in terms of accuracy, recognition percentage, and overall number of false positives [11].

A deep learning binomial classification for NIDS is suggested and successfully assessed by [12] utilizing the UNSW-NB15 database. The best activation operation, the most crucial characteristics, and the suggested model's testing on hypothetical

Table 1 Summary of the literature review

Techniques	Database	Factors of performance	Limits
RFs & SVMs	NSL-KDD	Accuracy, precision, recall	May used some more Classification techniques
DLs	NSL	Accuracy, precision, recall, F-Score, false alarm	Real-world network traffic Datasets
RNNs	KDDs KDDCUP 99	TPR, FPR, the detection rate	To reduce the training time using GPU acceleration
Evolutionary General RNNs	NSL-KDD	Accuracy, TPR, FPR detection rate, FNR	Need more trained dataset with feature reduction techniques
PSO with Gini index	UNB ISCX	Accuracy, detection rate, precision, F1-Score, and false alarm	Performance may increase using some machine learning model
Density Peak Clustering Techniques and DBNs	IDs database	Accuracy, detection rate and FNR	The aggressive training approach will be used to combine U2R and R2L attacks.
Multilayer feed forward ANNs using back propagation	NSL-KDD	Accuracy, false positive rate, precision, F1-Score, and recall	The performance will also measure in some other data types
DT approach	NSL-KDD	Accurate positive, false positive, and processing time	The performance will also measure in some other performance parameters value

data were all accomplished via the execution of three separate experiments. The model suggested exceeds previous frameworks in the research with 98.89% efficiency and 0.58% rate of false alarms on unknown information. And Table 1 summarizes the literature survey.

3 Proposed Methodology: Fuzzified Deep Convolutional Neural Network (FDCNN)

This technological study introduces a cross-layer security protocol that employs a fuzzy oriented trust calculating technique and utilizes trust to assess a node's appropriateness for the routing process. This estimate procedure may use several sensor node inputs, make appropriate judgments, and manage the unexpected to get the best outcome so that belief is not assumed to be the probability. The failure surveillance system is based on the ECNNs classifier algorithm, and trust threshold amounts are used to first detect invasive networks in the overall network prior to eradicate problem networks.

3.1 *Cross Layer Architecture Work on Fuzzy-Based Trusted Evaluation*

In this study, a wireless channel and fuzzy logic computation system create information for the suggested crossing layer method utilizing flexible logic confidence, enabling safe routing in WSN. It is realized in both the network acquisition stage (assessing eligible sensing networks' commencement) and the information forwarding stage (trust among suitable sensing networks). To begin the secure choosing of node sensors for transmitting and routing information packets to their final destination, this decision is made.

3.1.1 Stage on Network Acquiring

When another node "A" broadcasts the RTSs signal including the position of component "A"s and the target node within its transmission region, the data collecting procedure is launched. The connected network replies with a CTSs signal to network node "A" after receiving the broadcast. The fuzzy logic technique computes the applicability of contributing nodes that collect data as the number of contacts increases, recalculating and updating trust levels for each node. The fuzzy logic and static variables that are used to recreate the border criteria provide preference to the network if it displays the highest trust score.

3.1.2 The Data Forwarding Phase

The data forwarding step begins after selecting the right node. From the original network to the forwarded network and finally to the node with the destination, information is sent. The various conditions that include the achievement towards the percentile for sending data involve data transmission reliability, transfer internal (), how long it is taken for sending a packet of information in regard to the total amount of packets of data passed on inside the time structure, and equality ratio (fr), which decides and selects the relay network in next hop among adjacent nodes. A node is punished by having its fr and values decreased if it participates in the forwarding method before and is discovered to be doing so again. This is done to ensure that the infrastructure is divided for the subsequent phase and that disruptive behavior is prevented.

Equation 2, which is the evaluation of initiation (AI) when combined therewith the confidence (T) acquisition among the sensor node locations, develops a recommended cross-layer structure utilizing the idea of trust R.

AI is stated differently from its binary equivalent, i.e., it has a fuzzy value. Furthermore, SNR is denoted by RTS, which the tenacity of the packet for sending communication is from acquired the RTS. The frequency at which messages are being conveyed by a single node is indicated by relay and it is calculated by the period necessary for transmission.

T is used to update trust. During the transmission of information in the form of packets, the routing iterates *T*'s starting values. Additionally, the fairness ratio (fr), which selects the optimal route, denotes the fraction of successfully delivered packets (*r*), while the time of data transmission () is assessed separately. The discrepancies in Eqs. 3 and 4 produce the variable array threshold amounts. These requirements provide the cross-layering framework's ability to mitigate the effects of initiated security assaults while maintaining the resilience of interaction which defines the assessment of initiation using trust. The node's reputation is trusted if ($T <$ TTh), unsure if Tmin $< T <$ TTh, and feared when it drops below the threshold. This trust calculation makes use of the fuzzy logic trust estimate procedure to determine the trust value.

The Bayesian value of packets in WSNs which were introduced fraudulently and deleted the information packages, is referred to as the packets lost by nodes A and B that meet the criteria below.

1. Packets that Node A sends to Node B but that Node B discards.
2. The overall messages that node B has rejected, irrespective of whether they were caused by traffic.
3. How Node A determines Node B's precedence in relation to the other packets from the other Nodes.
4. Node B's failure to convey messages promptly is referred to as a malevolent node sending the messages, and it occurs under the following circumstances.

   - Packets that node A misrouted
   - Packets that node A deceptively inserted

By considering mimic actions, the congestion and quantity and variables needed to provide trust-based range to permit the functioning level of network that monitor that were shown as the proposed methodology in Fig. 1.

3.2 Regularizing Deep Neural Network Classifier

Deep neural networks include CNNs. CNN algorithm forward and back propagation are analyzed in the suggested extensive study. The theoretical maximum speed increase and parallel efficiency are calculated by measuring forward and reverse processing time. CNN is typically used to identify two-dimensional visuals that include displacement, zooming, and other unsettling invariances. Since CNN's feature recognition layer learns from training data, formal extracting features are not needed. Instead, CNN uses training data to do implicit training. Regarding the neural network that is interconnected, this is a significant advantage of the convolutional network. Its design is remarkably reminiscent to a biological brain network. Shared values aid in bringing down the network's size. Multi-dimensional output

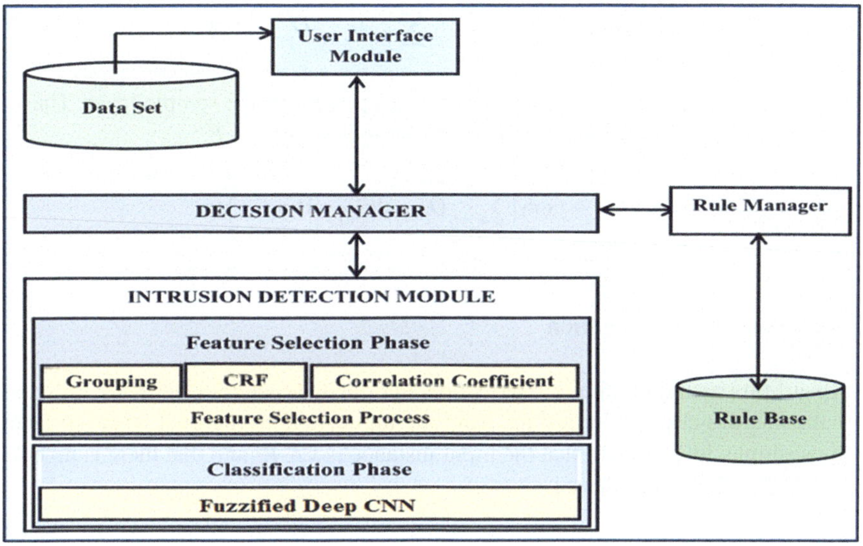

Fig. 1 Flow diagram of proposed methodology

vector graphics may immediately enter the machine learning system, which simplifies data reconstruction during extraction of features and classifications.

3.2.1 Back Propagation

The convolution and sampling functions of the CNN method are both crucial. A trainable filtering *Fx* is used, then deconvolution is performed of the original picture then a bias bx is included, and the convolution layer *Cx* is created. A single pixel is created during the procedure of sampling by combining n images from every neighborhood, which are then weighted using a scalar weighted $Wx + 1$, added with a bias of $bx + 1$, and formed as a restricted n times map of characteristics $Sx + 1$ using a function of activation. CNN's fundamental method uses the local receptive area, weight collaborating, along with time or space subsampling to extract features and decrease training factors.

$$O_{x,y}^{(l,k)} = \tanh \sum\nolimits_{t=0}^{f-1} \sum\nolimits_{r=0}^{Kh} \sum\nolimits_{c=0}^{Kw} W_{(r,c)}^{(k,t)}\, o_{(x+r,x+c)}^{(l-1,t)} + \mathrm{Bias}^{(l,k)} \tag{1}$$

f between all denotes the quantity of convolutional layers in a feature structure. The neuron's outputs for row *x*, column *y*, and the ith sub sampling level and k^{th} characteristic pattern are as follows:

$$O_{x,y}^{(l,k)} = \tanh(W^{k} \sum\nolimits_{r=0}^{S_h} \sum\nolimits_{c=0}^{Sw} O_{(x\times S_h+r,y\times Sw+c)}^{(l-1,t)} + \mathrm{Bias}^{(l,k)}. \tag{2}$$

$$O_{(i,j)} = \tanh\left(W^k \sum_{k=0}^{S-1} \sum_{x=0}^{S_h} \sum_{y=0}^{S_w} W_{(x,y)}^{(j,k)} O_{(x,y)}^{(l-1,t)} + \text{Bias}^{(lk)}\right) \tag{3}$$

S among these denotes how many features are present in the sample layer. The Ith neuron's outlet is represented by the Ith output region of level F.

$$O_{(i,j)} = \tanh\left(\sum_{j=0}^{H} O_{(l-1,j)} W_{(i,j)}^{l} + \text{Bias}^{(l,i)}\right) \tag{4}$$

3.2.2 Forward Propagation

Forward propagation computes and stores intermediate values (including outputs) for the neural networks in designs from the input stage to the final level. To keep things simple, let's assume that the input instance is $x \in R^d$ and that there is no bias term. The middle factor here is:

$$z = W^{(1)}{}_x \tag{5}$$

The buried layer's weight parameter is identified by the notation $W^{(1)} \in R^{h \times d}$. After the basic elements manipulate an intermediate variable $z \in R^h$ into the activating operate, an unseen layer ϕ variable with the vector's duration h is obtained.

$$h = \phi(z) \tag{6}$$

In addition, the hidden constant h is a middle variable. Obtain a layer output constant with a vector with a length of q, presuming that the resulting layer's variables only have a value of $W^{(2)} \in R^{q \times h}$:

$$o = W^{(2)}h \tag{7}$$

Where, l is the loss function, and y is the instance label, estimates the loss word for one information example,

$$L = l(o,y) \tag{8}$$

The concept of ℓ_2 norm periodicity states that given the hyperparameter λ, the term for normalization is

$$s = \frac{\lambda}{2}\left(\left\|W^{(1)}\right\|_F^2 + \left\|W^{(2)}\right\|_F^2\right) \tag{9}$$

$$J = L + s \tag{10}$$

Here, J serves as the information provided example's function of objective value.

4 Experimental Analysis

The entire system had been classified as multiple classes for every single assault, indicating the system's nth efficiency. Consequently, we have succeeded in achieving the following goals:

1. Identifying the important components which may aid in high-performance recognition
2. Applying the deep learning technique to attack recognition
3. Obtaining the highest level of efficiency in comparison to current solutions

4.1 Performance Measurements

The success of machine learning could be evaluated using a variety of metrics. In a research, we examined the performance indicators including specificity, sensitivity, reliability, recall, *F*-score statistics, and ROC Curves.

$$\text{Accuracy} = \frac{\text{TPs} + \text{TNs}}{\text{TPs} + \text{FPs} + \text{FNs} + \text{TNs}}^{*} 100\%$$

$$\text{Sensitivity} = \frac{\text{TPs}}{\text{TPs} + \text{FNs}}^{*} 100\%$$

$$\text{Percision} = \frac{\text{TPs}}{\text{TPs} + \text{FPs}}^{*} 100\%$$

$$F\text{-score} = \frac{2^{*}\,\text{Precision}^{*}\,\text{sensitivity}}{\text{Percision} + \text{Sensitivity}}^{*} 100\%$$

For data misidentified as an assault, FPs or TNs were recorded. TPs are information which has been correctly recognized as a malicious attempt. FNs were indications when an attack information set was incorrectly classified as neutral. FDCNN were performed well in all datasets with accuracy scores of 58–66%. UDP assaults were 100% detected by the FDCNN classifier.

4.2 Discussion

The fourth agricultural revolution combines cutting-edge technological innovation with traditional farming. Security professionals are now researching a variety of cyber threats connected to this connectivity. Agriculture 4.0 is a new era of agriculture brought about by fast technological breakthroughs. Agriculture 4.0 uses new technology and methods to solve current issues, reduce risks, and improve productivity due to climate change, diseases, overuse of pesticides and assets, and more. This is Agriculture 4.0's objective. This goal is achieved by using a wide range of ICTs that are state-of-the-art in their respective industries. Developing Agriculture 4.0 is going to be most difficult since thousands of IoT-based devices will be placed in open spaces. Safety and privacy are expected to pose the biggest challenges. Because of this, each layer in the framework of the IoTs has its own unique set of privacy and security issues [13]. DDoS attacks, which are among the most common cyberattacks, have a chance to jeopardize the security of food, the efficiency of agri-food distribution networks, and the productivity of farms [14].

We built a system that utilized the LSTM and used CNN–LSTMs algorithms to identify 4.0 agricultural attacks. The CIC-DDoS2019 dataset was used in order to assess the suggested system [15]. The FDCNNs algorithm was extremely accurate when it comes to classifying a wide range of categories. The FDCNNs algorithm's excellent performance as a whole yielded a flawless score of 100% when tested against various types of attacks [16].

Regarding the recall measure of multiclass categorization for the purpose of recognizing interference, it was discovered in Agriculture 4.0 that the FDCNN model outperformed other approaches [17]. The recollection for categorizing the five attacks is shown on the y-axis, while sensitivity measurements are portrayed on the x-axis, in the ROC measure utilized to verify the outcomes of the LSTM and FDCNNs techniques. It was found that this FDCNNs model were effective in reaching highest value of 100% for recognizing five forms of assault in addition to the typical one as shown in Fig. 2.

The proposed solution DL CNN–LSTMs method to the method's extant security features was developed with the available data. Figure 3 compares our method's efficiency numbers with those of other methods. The technique we recommend has the best level of accuracy overall compared to the ones which are presently available. Table 2 displays the comparison of the FDCNN model with current security measures to find threats on Agriculture 4.0, and Fig. 4 compares the proposed model's efficacy.

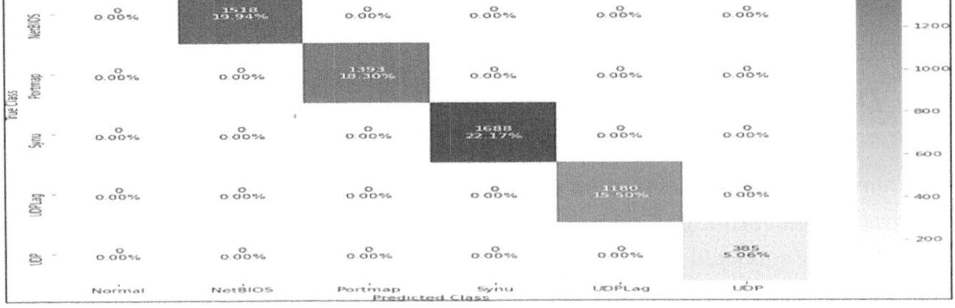

Fig. 2 The confusion matrix of proposed FDCNN

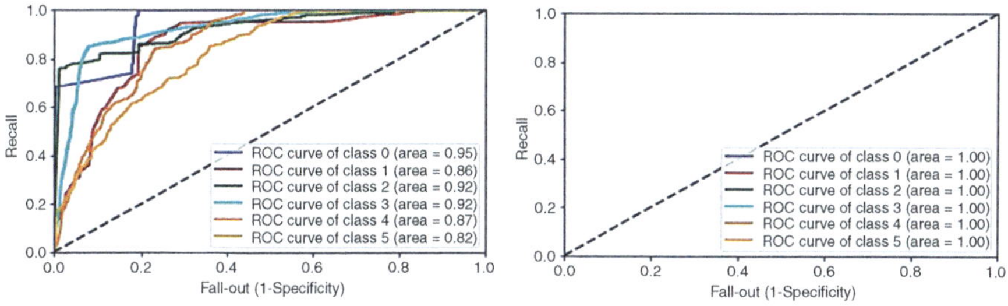

Fig. 3 ROC of suggested system: (**a**) DCNN system; (**b**) FDCNN system

Table 2 Results of comparing the FDCNN model with current security measures to find threats on Agriculture 4.0

Techniques	Database	Factors of performance	Limits
DCNNs	CIC-DDoS2019	Multi-classifier	99.8%
CNNs			96.5%
Boost Techniques			94.28%
FDCNN			100%

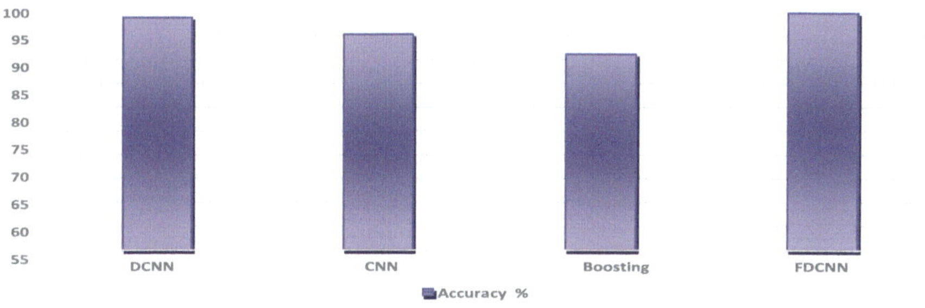

Fig. 4 Comparison of the proposed model's efficacy

5 Conclusion

In conclusion, the Fuzzified Deep Convolutional Neural Network (FDCNN) presents a novel and effective approach for network attack detection. By integrating fuzzy logic and deep learning techniques, the FDCNN surpasses traditional methods in terms of accuracy, robustness, and real-world applicability. Through the fusion of fuzzy logic and deep learning, the FDCNN enables the representation of uncertain and imprecise information associated with network traffic data. This fuzzification process captures the inherent complexity and variability of network attacks, leading to improved detection performance. The deep convolutional neural network architecture of the FDCNN facilitates the learning of hierarchical and abstract representations of network traffic patterns. By automatically extracting discriminative features from raw network data, the FDCNN achieves superior accuracy and reduces false positives and false negatives in network attack detection. The practical applicability of the FDCNN is evident in its potential for deployment in real-world network security systems. By effectively detecting network attacks, it contributes to enhancing the security and resilience of computer networks across diverse fields and applications. Additionally, the Fuzzified Deep Convolutional Neural Network (FDCNN)-based IDS model outperformed more recent deep learning-based IDS algorithms in terms of performance. With the use of the CIC-DDoS2019 dataset, several methods were evaluated. The outcomes demonstrated that the FDCNN model completely succeeded in identifying network threats. Moreover, the FDCNN-based IDS model demonstrated noticeably better performance when compared to other cutting-edge deep learning IDS methods.

Further research can focus on optimizing the FDCNN model and exploring its application in different network environments to validate its efficacy and adaptability. The continued development of such advanced techniques is crucial for ensuring the integrity and security of computer networks in the face of evolving threats.

Conflict of Interest The authors declare that they have no conflict of interest.

Ethics Statement This chapter does not contain any studies with human or animal subjects performed by any authors. Informed consent was obtained from all individual participants included in the study.

Data Availability Statement Available Based on Request. The datasets generated and/or analyzed during the current study are not publicly available due to the extension of the submitted research work but are available from the corresponding author on reasonable request.

References

1. Das, R., Sen, S., & Maulik, U. (2020). A survey on fuzzy deep neural networks. ACM Computing Surveys (CSUR), 53(3), 1–25.
2. Liu, M., Zhou, Z., Shang, P., & Xu, D. (2019). Fuzzified image enhancement for deep learning in iris recognition. IEEE Transactions on Fuzzy Systems, 28(1), 92–99.
3. GSR, E. S., Azees, M., Vinodkumar, C. R., & Parthasarathy, G. (2022). Hybrid optimization enabled deep learning technique for multi-level intrusion detection. Advances in Engineering Software, 173, 103197.
4. Javaheri, D., Gorgin, S., Lee, J. A., & Masdari, M. (2023). Fuzzy logic-based DDoS attacks and network traffic anomaly detection methods: Classification, overview, and future perspectives. Information Sciences.
5. Mishra, P., Varadharajan, V., Tupakula, U., & Pilli, E. S. (2018). A detailed investigation and analysis of using machine learning techniques for intrusion detection. IEEE communications surveys & tutorials, 21(1), 686–728.
6. Deepika, J., Rajan, C., & Senthil, T. (2021). Security and privacy of cloud-and IoT-based medical image diagnosis using fuzzy convolutional neural network. Computational Intelligence and Neuroscience, 2021, 1–17.
7. Talpur, N., Abdulkadir, S. J., Alhussian, H., Hasan, M. H., Aziz, N., & Bamhdi, A. (2023). Deep Neuro-Fuzzy System application trends, challenges, and future perspectives: A systematic survey. Artificial intelligence review, 56(2), 865–913.

8. Kishore, R., & Chauhan, A. (2020, November). Evaluation of deep neural networks for advanced intrusion detection systems. In 2020 4th International Conference on Electronics, Communication and Aerospace Technology (ICECA) (pp. 1–8). IEEE.
9. Ge, C., Liu, Z., Fang, L., Ling, H., Zhang, A., & Yin, C. (2020). A hybrid fuzzy convolutional neural network based mechanism for photovoltaic cell defect detection with electroluminescence images. IEEE Transactions on Parallel and Distributed Systems, 32(7), 1653–1664.
10. Novaes, M. P., Carvalho, L. F., Lloret, J., & Proenca, M. L. (2020). Long short-term memory and fuzzy logic for anomaly detection and mitigation in software-defined network environment. IEEE Access, 8, 83765–83781.
11. Mendis, G. J., Kamal, M. B., & Wei, J. (2019). Intelligent situational-awareness architecture for hybrid emergency power systems in more electric aircraft. Deep learning applications for cyber security, 27–44.
12. Zheng, Y., Xu, Z., & Wang, X. (2021). The fusion of deep learning and fuzzy systems: A state-of-the-art survey. IEEE Transactions on Fuzzy Systems, 30(8), 2783–2799.
13. Ding, W., Abdel-Basset, M., Hawash, H., & Moustafa, N. (2022). Interval type-2 fuzzy temporal convolutional autoencoder for gait-based human identification and authentication. Information Sciences, 597, 144–165.
14. Bendre, N., Ebadi, N., Prevost, J. J., & Najafirad, P. (2020). Human action performance using deep neuro-fuzzy recurrent attention model. IEEE Access, 8, 57749–57761.
15. Buczak, A. L., & Guven, E. (2015). A survey of data mining and machine learning methods for cyber security intrusion detection. IEEE Communications surveys & tutorials, 18(2), 1153–1176.
16. GSR, E. S., Ganeshan, R., Jingle, I. D. J., & Ananth, J. P. (2023). FACVO-DNFN: Deep learning-based feature fusion and Distributed Denial of Service attack detection in cloud computing. Knowledge-Based Systems, 261, 110132.
17. Li, F. Q., Wang, S. L., Liew, A. W. C., Ding, W., & Liu, G. S. (2020). Large-scale malicious software classification with fuzzified features and boosted fuzzy random forest. IEEE Transactions on Fuzzy Systems, 29(11), 3205–3218.

Part II
Big Data Analytics and Machine Learning

Evolvable Hardware Using Genetic Algorithm

B. Gomathi, R. Manimegalai, S. K. Noor Mohamamd, G. V. Sri Rajiv Jegan, and M. Venkateshwaran

1 Introduction

1.1 Evolvable Hardware

Evolvable Hardware (EH) focuses on employing evolutionary algorithms to design specialized electronic devices without the need for human expertise [1, 2]. It combines fault tolerance, autonomous systems, evolutionary computation, and reconfigurable hardware. Hardware can adapt to its environment and modify its architecture and behavior on the fly is referred to as Evolvable Hardware. This provides applications with unprecedented levels of flexibility and adaptability, and the potential for drastic transformation in the development and utilization of electronic systems is held by it. The evolvable hardware is used across a wide range of industries such as telecommunications, aircraft, and robotics. Hence, the idea of evolvable hardware and the most recent advancements in the industry are examined in this work. Evolvable Hardware may be used to build and optimize robotic systems that can adapt to changing environmental circumstances and perform a variety of jobs in the field of robotics, which is one important application. It may also be utilized to create reconfigurable computer systems, in which the hardware can be altered to enhance the performance of specific applications. It may also be used in the field of fault-tolerant computing to create systems that can continue to function in the face of hardware failures. In addition to the aforementioned uses, evolving

B. Gomathi (✉) · R. Manimegalai · G. V. Sri Rajiv Jegan · M. Venkateshwaran
PSG Institute of Technology and Applied Research, Coimbatore, India
e-mail: drrm@psgitech.ac.in; 19cs136@psgitech.ac.in

S. K. Noor Mohamamd
Indian Institute of Information Technology, Design and Manufacturing, Kancheepuram, India

A. Haldorai et al. (eds.), *6th EAI International Conference on Big Data Innovation for Sustainable Cognitive Computing*, EAI/Springer Innovations in Communication and Computing, https://doi.org/10.1007/978-3-031-54696-9_5

hardware can also be used to build adaptive and intelligent systems. In the area of signal processing, it may be used to enhance the performance of communication systems such as wireless networks and satellite communication systems. Evolvable Hardware may be used to create hardware-based security systems, where the hardware can be altered and improved to offer greater defense against security threats and cyber-attacks. Due to its ability to adapt and evolve, Evolvable Hardware is often a beneficial tool for improving and optimizing a variety of computer systems and applications. In this work, Field Programmable Gate Arrays (FPGAs) is used to design and assess evolvable hardware for digital circuits. To get around the hardware failure, GA operators such as crossover, mutation, and selection operators are used to alter the dynamic bit stream.

1.2 Evolutionary Algorithm

Evolutionary Algorithms (EAs) are optimization techniques that are inspired by biological evolution. They are excellent at resolving challenging issues that traditional optimization techniques find hard or unattainable. Evolutionary algorithms iteratively develop a population of viable solutions over several generations by imitating the processes of natural selection, mutation, and reproduction [10, 11]. This repetitive process enables EAs to investigate broad solution spaces, adapt to shifting surroundings, and converge towards ideal or nearly ideal solutions for difficult situations that challenge conventional methods. Typical phases of an evolutionary algorithm include the following phases:

1. Initialization: The population of potential solutions is produced at random.
2. Selection: A subset of the population is chosen to serve as the parents for a fresh batch of candidate solutions. Candidates are typically selected based on their fitness scores.
3. Reproduction: By recombination i.e. crossover and mutation operators, the chosen parents are employed to create a new population of candidate solutions.
4. Termination: After certain amount of generations or evaluations, the algorithm comes to an end when a good solution is identified.

EAs may be employed to address a variety of issues, such as design, machine learning, and optimization. They have been used to solve issues including portfolio optimization, parameter adjustment for machine learning models, and the creation of intricate systems such as antennas and airplane wings. EAs are widely utilized in several practical fields, including engineering, finance, and logistics. EAs can be used to solve evolvable hardware problem as shown in Fig. 1.

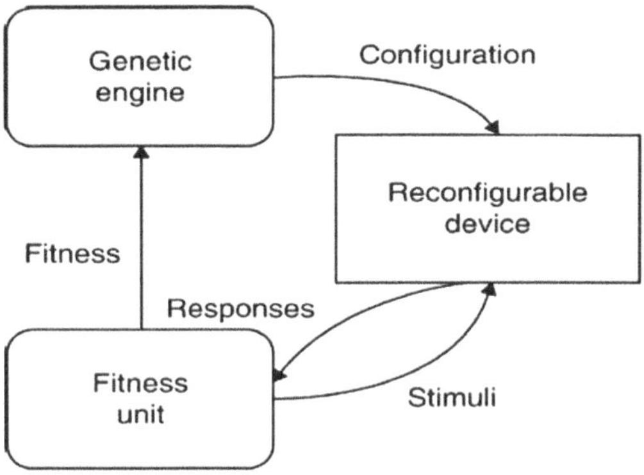

Fig. 1 Use of Evolutionary Algorithm in Evolvable Hardware

1.3 Genetic Algorithm

GA is a type of Evolutionary Algorithm (EA) that is based on the principles of natural selection and genetics. Using genetic operators including selection, crossover, and mutation, GA builds a population of potential solutions, evaluate their fitness, and iteratively produce new solutions. This procedure continues until the best solution is discovered or the termination requirements is satisfied. GAs effectively explore large solution spaces by maintaining a balance between exploration and exploitation, making them essential to effectively resolving challenging problems. The key components of a GA are:

1. Fitness function: Each potential solution is graded on its ability to solve the problem using a fitness function.
2. Selection: The population's fittest members are selected to be the parents of the generations that follow.
3. Crossover: The selected parents are recombined to generate new offspring.
4. Mutation: The offsprings are subjected to random mutations to introduce new genetic variation.
5. Termination: When a workable solution is identified, the GA terminates, or when a certain amount of generations or evaluations are completed. Field-Programmable Gate Arrays (FPGAs), digital circuits, and other forms of hardware are evolved using evolutionary algorithms in evolvable hardware. The advantage of evolutionary algorithms in evolvable hardware is their ability to automate design processes, explore complex design spaces, adapt to changing requirements, and find novel and optimized hardware solutions, making them valuable tools in contemporary engineering and technology. Hence, the Genetic Algorithm is used in the proposed work to solve the evolvable hardware problem and produce the expected result.

2 Literature Survey

Abishek et al. have proposed a Genetic Algorithm to design evolvable hardware [3]. It is mainly used in the field of Artificial Intelligence. While comparing to previously generated solutions, the Genetic Algorithm provides the best solution by using the basic concepts of genetic operators such as crossover, mutation, etc. It helps them by generating the best solution for the problem and reducing the time of execution. It also solves more complex problems very easily. Genetic Algorithm is reviewed in more detail in the following section, which also gives a general overview of the computer model's ideas. The proposed genetic algorithm mainly depends upon random experiments of reproduction. This work helps us to solve the promising approach towards autonomous and online reconfigurable machines capable of adapting and providing the best approach to real-world problems. Stomeo et al. have proposed a brand-new Genetic Algorithm for hardware evolution [4]. Using evolutionary algorithms, search and optimization issues can be solved. However, due to scalability, evolvable hardware is not frequently used to address applications in the real world. The work uncovers a novel evolutionary algorithm that is tested for scalability and is partially designed to solve logic gates. The approach is based on the Programmable Logic Array (PLA). Without using any decomposition techniques, it aids in the growth of huge logic circuits. Based on several logic circuits, this method generates results. This algorithm provides a better and optimal solution.

Krishnaveni [5] et al. have proposed an evolvable hardware architecture using a Genetic Algorithm for distributed arithmetic FIR filter. The distributed arithmetic approach is used to reduce the complexity of the filter implementation, and the optimization of the filter coefficients is done using a Genetic Algorithm. The proposed architecture is tested on several benchmark signals and compared with other filter architectures. The proposed algorithm consists of two stages, namely, initialization and evolution. In the initialization stage, a population of filter coefficients is randomly generated. In the evolution stage, the general population of coefficients is evolved via the genetic method. The mean squared difference between the desired output and the filter output is the fitness function. The best filter coefficients are selected as the final solution. The proposed architecture reduces the complexity of the filter implementation using the distributed arithmetic approach. The genetic algorithm optimizes the filter coefficients to improve the filter performance. The proposed architecture is tested on several benchmark signals and outperforms other filter architectures in terms of accuracy and computational complexity. Krishnaveni [6] et al. have presented a novel approach for designing a distributed arithmetic Finite Impulse Response (FIR) filter using an evolvable hardware architecture and a Genetic Algorithm. The proposed evolvable hardware architecture is described for implementing the filter using Field Programmable Gate Arrays (FPGAs). The architecture includes a genetic algorithm-based reconfigurable module that can be used to optimize the filter coefficients for better performance. The results show that the proposed approach outperforms the conventional FIR filter and other existing approaches in terms of area, speed, and power consumption. Overall, the chapter provides a significant contribution to the field of evolvable hardware and digital signal processing. The use of genetic algorithms for optimizing the filter coefficients and the evolvable hardware architecture for implementing the filter can lead to improved performance and reduced hardware complexity. Sim and Harashiam have proposed the design of evolvable hardware which is based on Genetic Algorithm Processor (GAP) [7]. The GAP is a hardware accelerator designed specifically for genetic algorithms, which allows for the parallel evaluation of multiple candidate solutions. It is a combination of genetic algorithm with digital hardware in order to create self-evolving electronic circuits. This approach is based on the concept of using genetic algorithm to optimize the performance of a digital hardware circuit, which can then be reconfigured and improved over time. The GAP algorithm is specifically designed to optimize the performance of digital hardware circuits. This algorithm produces the population for candidate circuits; each population is a set of binary numbers. These circuits examine the fitness function which calculates the performance of a specific task. The fittest circuits are then selected for reproduction, where they are combined with other circuits to create new offspring circuits. This procedure is repeated over multiple generations, gradually improving the performance of the circuit. The advantages of evolvable hardware based on GAP include the ability to create highly optimized digital hardware circuits that are specifically designed for a particular task or application. This approach allows creation of circuits that can adapt and evolve over time, improving their performance. Additionally, evolvable hardware based on GAP can be used to create circuits that are more robust and reliable, as they can adapt to changes in their environment or operating conditions.

A novel approach is proposed to design evolvable hardware using a Genetic Algorithm Processor (GAP) [8, 9]. Design of hardware systems that can adapt and evolve over time in response to changing environmental conditions or user requirements is discussed in [8, 9]. The architecture and operation of the GAP is presented in detail, highlighting its ability to perform fitness evaluation, selection, and reproduction operations in hardware. A case study that uses the GAP to evolve a 16-bit multiplier circuit is presented.

3 Genetic Algorithm in Evolvable Hardware

Digital circuits, Field-Programmable Gate Arrays (FPGAs), and other hardware components are significantly improved and evolved using Genetic Algorithms (GAs). In this work, GA-driven intrinsic repair is proposed for FPGA-based digital systems. A significant evolution speedup is provided by the proposed work, which leads to improved system availability. However, the automated exploration of intricate design spaces is made possible by the use of Genetic Algorithm. Evolvable hardware enables the identification of optimum hardware configuration that could be difficult to accomplish through manual design methods. The flow diagram of the Genetic Algorithm is shown in Fig. 2.

The steps involved in the proposed GA based approach for evolvable hardware are listed below:

1. Set up an initial random population.
2. Use a fitness function to assess every individual.
3. Depending on user options, the individuals are chosen for reproduction using a roulette wheel, tournament, or rank-based selection mechanism. An individual may be able to opt for reproduction if it has the best fitness score.
4. Use crossover and mutation to facilitate reproduction, then calculate each individual's fitness.
5. Choose the best individuals based on fitness value for the next generation.
6. Proceed to Step 4 if the termination condition is not met. Otherwise, generate the optimal solution as an output. This optimal solution is used to evolve the digital circuit.

The Genetic Algorithm starts with initialization. Through this process, the initial population is produced which is a group of chromosomes. A set of parameters

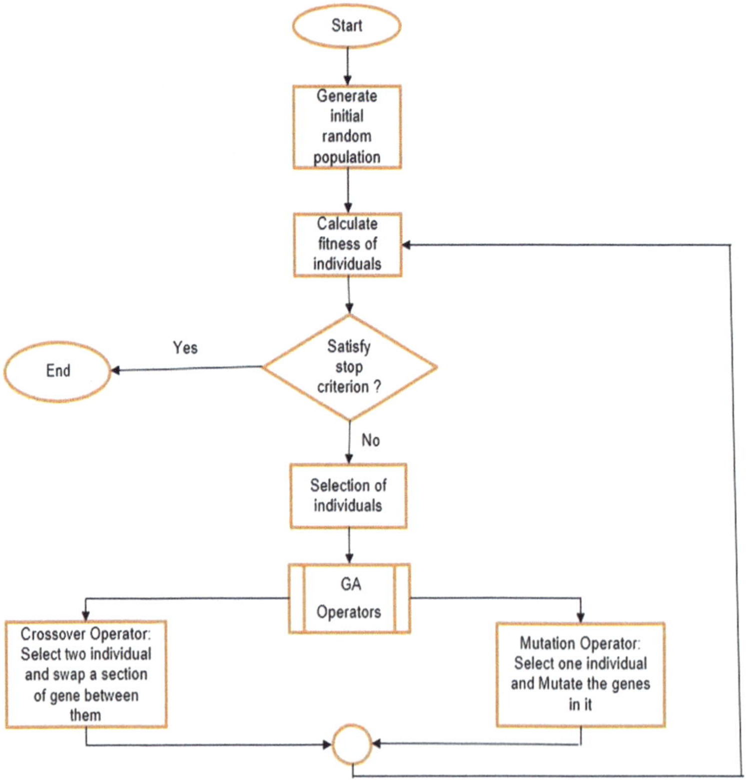

Fig. 2 Steps Involved in the Implementation of Genetic Algorithm

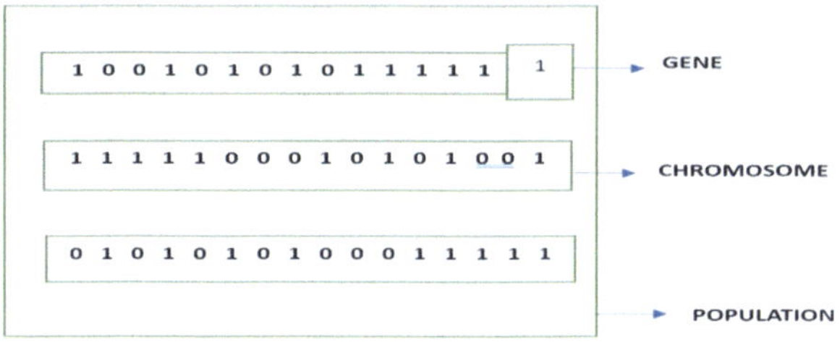

Fig. 3 Chromosome Representation

Fig. 4 Choosing an Offspring using Selection Operator

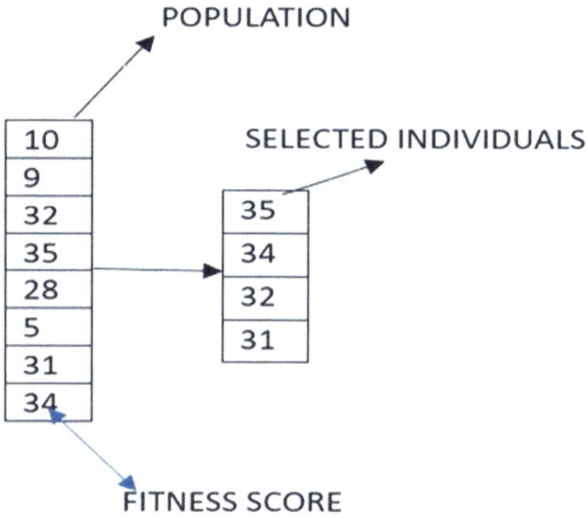

known as genes make an individual. Chromosomes are created from a string of genes as shown in Fig. 3.

Fitness function is computed to determine an individual's ability to compete with other people. Based on their degree of fitness, individuals are evaluated and a score is assigned. This score is used to determine the selection for reproduction. If an individual has the highest fitness score, then it has a possible chance of reproduction. The selection step involves choosing an individual to reproduce an offspring as shown in Fig. 4.

All the selected individuals are paired off or organized into orders to aid in reproduction. The subsequent generation is then affected by these individuals' genes. Three categories of choice exist:

- Roulette wheel selection: Based on its fitness, each prospective solution is given a probability of being chosen as shown in Fig. 5.
- Tournament selection: The population is divided into a random subset of individuals. The individual with the best level of fitness among those participating in the competition is chosen to be a parent for the following generation as shown in Fig. 6.

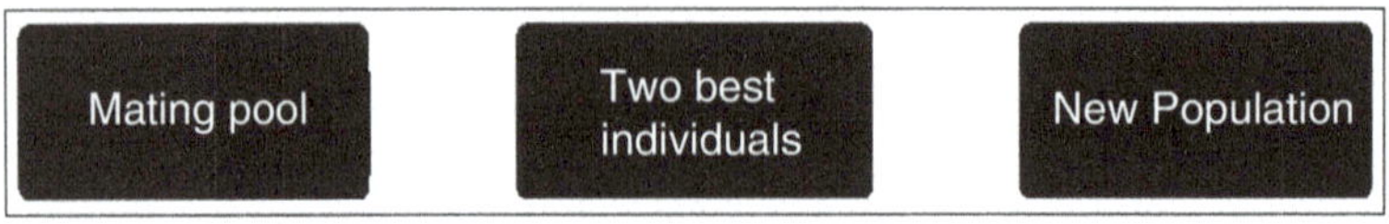

Fig. 5 Roulette-Wheel Selection

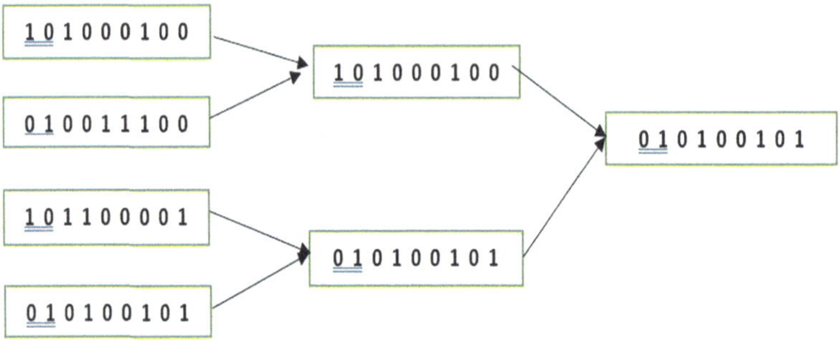

Fig. 6 Tournament Selection

Fig. 7 Crossover Operation

- Rank-based selection: Rank-based selection gives every individual a numerical rank depending on the fitness when compared to other individuals in the population.

In the reproduction process, a new individual is created. The Mutation and Crossover operations are applied to the parent population to generate new offspring. In this procedure, a crossover point is randomly selected within the genes. The crossover process then exchanges the genetic information of both parents from the present generation to create new individuals as shown in Fig. 7. Different crossover styles, such as single-point crossover and two-point crossover, are possible.

The mutation operator adds arbitrary genes to generate new offsprings to maintain population variety. Mutation is achieved by flipping a few chromosomal bits as shown in Fig. 8. It aids in addressing premature convergence as an issue and enhances diversification. It includes several types of mutation such as flip-bit mutation, interchanging mutation, and reverse mutation.

After the reproduction phase is completed, the termination condition is checked. The above steps are repeated until either an optimal solution with a threshold of fitness is attained or the genetic algorithm has gone through all of its possible iterations.

Fig. 8 Mutation Operation

Table 1 Parameters of Genetic Algorithm

Parameters	Value
Population Size	100
Chromosome Length	36
Number of Generations	30
Selection Rate	40
Crossover Points	2, 4, 8

4 Results and Discussions

The input parameters for the proposed GA based approach are listed in Table 1. A subset of the current generation's solutions is the population. Chromosome length is the number of bits in one single chromosome. The number of generations is the number of iterations that the genetic algorithm function is being called. Selection Rate is a parameter that defines how many numbers of individuals should be selected and passed on to the next generation.

The convergence graphs for various operators are shown in Figs. 9, 10, and 11. Though several offsprings result from the eight-point crossover with uniform mutation, there is a chance that specific patterns in the genetic structure are disturbed. Finding a balance between investigating novel combinations and making use of pre-existing genetic structures is achieved by the four-point crossover with uniform mutation. Therefore, when compared to two-point and eight-point crossovers, four-point crossover converges more quickly. Depending on the particular problem being handled, the effectiveness of crossover and mutation operators can vary.

Figure 12 presents the computational effort required to generate optimal solution using tournament selection and rank-based selection. It can be observed that rank-based selection performs better.

5 Conclusions and Future Work

In this work, Genetic Algorithm is applied to create evolvable hardware. Industrial applications can be implemented by using evolvable hardware, especially those that call for the solution of time-varying issues and real-time limitations. The chromosome representation of the circuit is crucial because complex fitness functions require a large number of architectural bits, which has an immediate impact on the

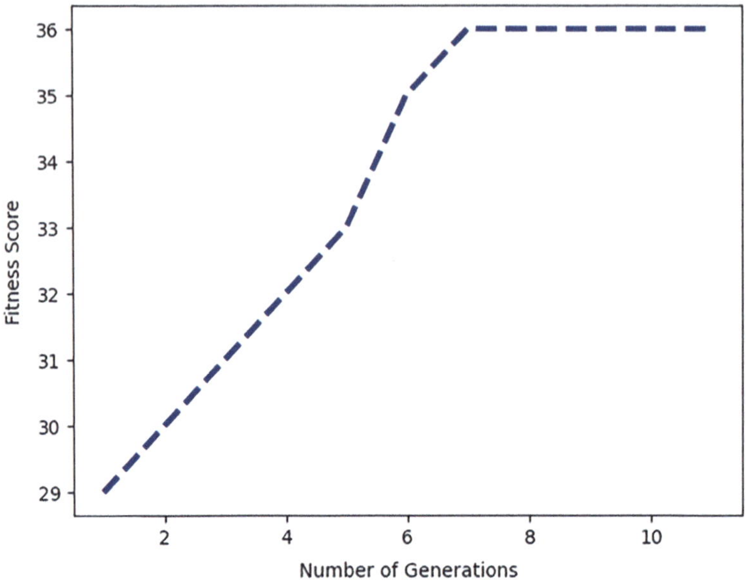

Fig. 9 Convergence Rate for Rank-based Selection with Two-Point Crossover and Uniform Mutation

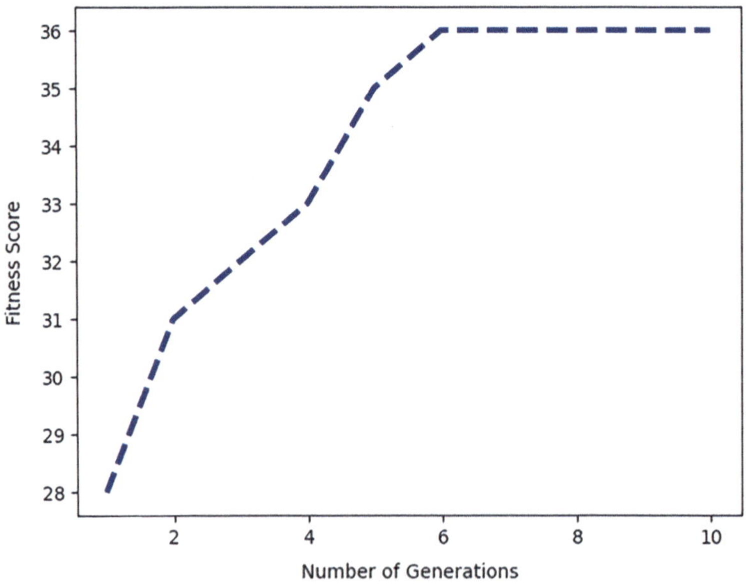

Fig. 10 Convergence Rate for Rank-based Selection with Four-Point Crossover and Uniform Mutation

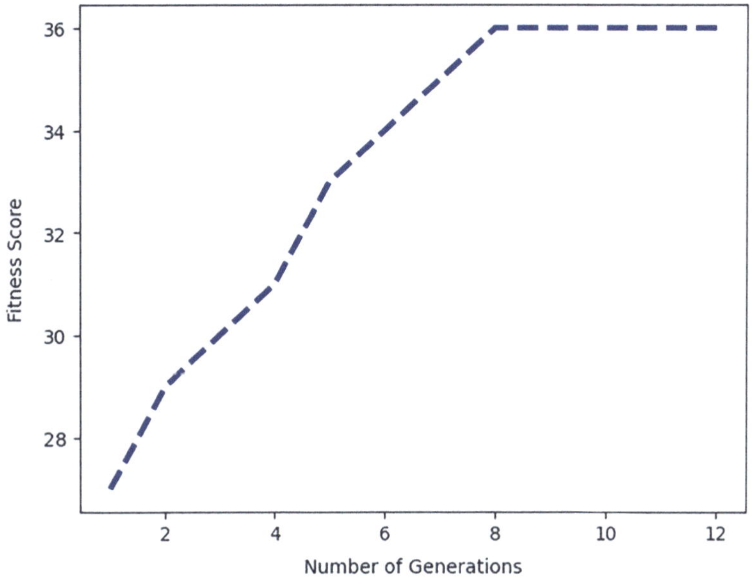

Fig. 11 Convergence Rate for Rank-based Selection with Eight-Point Crossover and Uniform Mutation

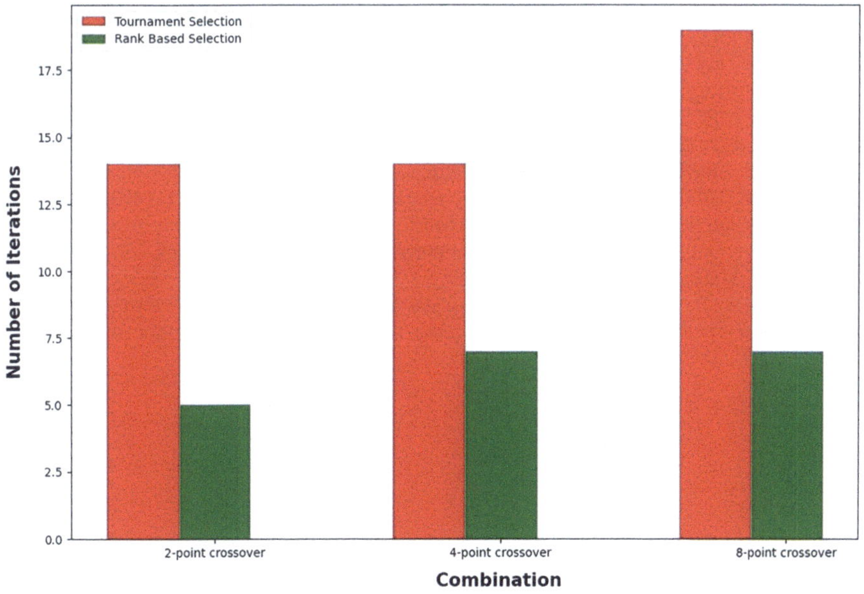

Fig. 12 Comparison between Tournament Selection and Rank-Based Selection Techniques

GA search space. Evolvable hardware can be successful when fitness approaches 100%, which may not always be attainable in large solution space. In this work, Genetic Algorithm is applied to evolvable hardware. The use of bio-inspired algorithms such as Particle Swarm Optimization (PSO) to design an evolvable hardware is an interesting open problem.

References

1. X. Yao and T. Higuchi, "Promises and challenges of evolvable hardware," In IEEE Transactions on Systems, Man, and Cybernetics, Vol. 29, No. 1, pp. 87–97, 1999.
2. J. R. Koza, M. A. Keane and M. J. Streeter, "Routine high-return human competitive evolvable hardware", In NASA/DoD Conference on Evolvable Hardware, USA, pp. 3–17, 2004.
3. Joglekar, Abhishek & Tungare, Manas, "Genetic Algorithms and their use in the Design of Evolvable Hardware", Advanced Engineering Informatics – Artificial, Vol.19. pp. 43–53, 2003.
4. E. Stomeo, T. Kalganova and C. Lambert, "A Novel Genetic Algorithm for Evolvable Hardware", IEEE International Conference on Evolutionary Computation, Canada, pp. 134–141, 2006.
5. Krishnaveni, K., Chandran Ranjith and S. P. Joy Vasantha Rani. "Evolvable Hardware Architecture Using Genetic Algorithm for Distributed Arithmetic FIR Filter. 2017.
6. Krishnaveni, K., Ranjith, C., Joy Vasantha Rani, S.P., "Evolvable Hardware Architecture Using Genetic Algorithm for Distributed Arithmetic FIR Filter", In: Dash, S., Vijayakumar, K., Panigrahi, B., Das, S. (eds) Artificial Intelligence and Evolutionary Computations in Engineering Systems. Advances in Intelligent Systems and Computing, Vol. 517, 2017.
7. K.-B. Sim and F. Harashiam, "Design of Evolvable Hardware based on Genetic Algorithm Processor(GAP)," International Journal of Fuzzy Logic and Intelligent Systems, Vol. 5, No. 3, pp. 206–215, 2005.
8. Ranjith, C., Joy Vasantha Rani, S.P., Priyadharsheni, B., Medhuna Suresh and Madhusudhanan, M.: Optimizing GA operators for System Evolution of Evolvable Embedded Hardware on Virtex 6 FPGA. ARPN Journal of Engineering and Applied Sciences, Vol. 10, No. 11, pp. 4908–4914, 2 015.
9. Ruben Salvador, Lukas Sekanina, Self-Reconfigurable Evolvable Hardware system for Adaptive Image Processing. IEEE transactions on computers vol. 62, No. 8, pp. 1481–1493, 2013.
10. Rashad S. Oreifej, Ronald F. DeMara, "Intrinsic evolvable hardware platform for digital circuit design and repair using genetic algorithms", Applied Soft Computing, Vol. 12, No. 8, pp. 2470–2480, 2012.
11. Gomathi, B., S. T. Suganthi, T. N. Prabhu, and Andriy Kovalenko. Multi-Objective Immune Algorithm for Internet of Vehicles for Data Offloading, Intelligent Automation & Soft Computing, Vol.34, No. 3, 2022.

Zero Watermarking and Data Authentication

R. Manimegalai, V. Megadarshini, SK Ayesha, and B. Gomathi

1 Introduction

In our modern digital landscape, security has ascended to the forefront of concerns within the realms of information technology and networked multimedia. As these domains continue to evolve at a relentless pace, the proliferation of digital media has created an urgent need for robust protection mechanisms against unauthorized access, reproduction, and distribution. This imperative to safeguard intellectual property has propelled the exploration of innovative and efficient security techniques to preserve the integrity of multimedia content [1, 2]. Security measures can be broadly classified into various categories, each tailored to address specific facets of safeguarding digital assets. Within this array of security methods, watermarking emerges as a prominent and versatile approach to secure multimedia content [3, 4].

Figure 1 illustrates the classification diagram in which watermarking is a pivotal component of data hiding techniques. Within the realm of security tools for safeguarding digital content, there are two primary categories, namely, content verification tools and content protection tools. Content verification tools are focused on authenticating and ensuring the integrity of digital content. Content protection tools, on the other hand, are designed to prevent unauthorized access, replication, or distribution of digital content. This category is further divided into encryption techniques for encoding content to preserve confidentiality and data hiding techniques

R. Manimegalai (✉) · V. Megadarshini · B. Gomathi
PSG Institute of Technology and Applied Research, Coimbatore, India
e-mail: drrm@psgitech.ac.in; d20z202@psgitech.ac.in

S. K. Ayesha
Vellore Institute of Technology, Chennai, India
e-mail: ayesha.sk@vit.ac.in

A. Haldorai et al. (eds.), *6th EAI International Conference on Big Data Innovation for Sustainable Cognitive Computing*, EAI/Springer Innovations in Communication and Computing, https://doi.org/10.1007/978-3-031-54696-9_6

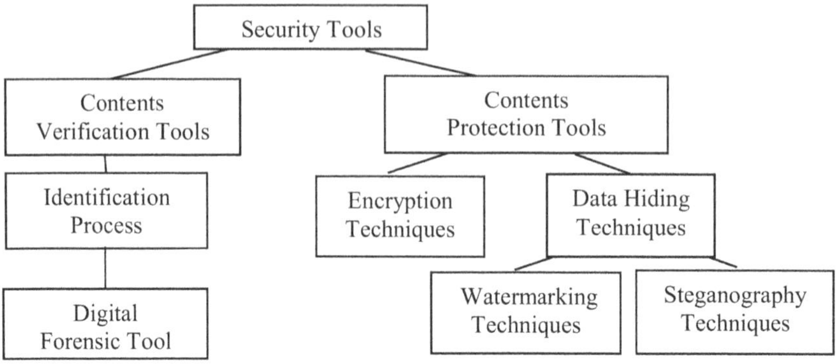

Fig. 1 Classification of Security Tools

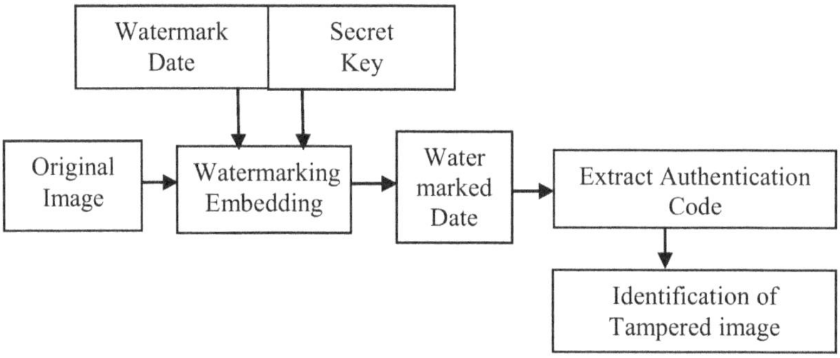

Fig. 2 Steps Involved in Watermarking

for concealing information within content. Data hiding techniques, in turn, have two subcategories: steganography techniques, which invisibly embed data within content, and watermarking techniques, which is used for copyright protection. In contents verification, the identification process involves transformations, often based on mathematical algorithms, to verify content authenticity. This hierarchical classification provides a structured framework for understanding and implementing a wide range of digital content security measures, from verification to protection and data hiding.

Watermarking, a specialized form of digital content protection, revolves around the discrete embedding of information within multimedia files as shown in Fig. 2. This embedded data, known as a watermark, functions as a digital fingerprint, carrying essential information such as copyright details, ownership information, or content authentication codes [5, 6]. Watermarking protects multimedia content against piracy while preserving the aesthetic or auditory qualities of the media. Watermarking is imperceptible to human observers. Watermarking carries several distinct advantages over other security methods. It excels in imperceptibility, ensuring that the presence of watermarks does not detract from the user's experience of the multimedia content. Additionally, it showcases robustness by withstanding various forms of modification or attacks, such as image compression, cropping, scaling, or minor alterations, while remaining detectable through appropriate decoding techniques.

Frequency-domain and spatial-domain watermarking techniques are the primary groups of modern digital image watermarking methods. Frequency-domain watermarking approaches are shown to be more successful in satisfying various requirements such as robustness and imperceptibility of digital watermarking algorithms than spatial domain techniques [7]. In this proposed work, a watermarking algorithm is combined with either Discrete Cosine Transform (DCT) or Discrete Wavelet Transform (DWT) in the transform domain. The image has all the data that reflects the characteristics of the original image because the information with low frequency is incorporated in DCT-based watermarking. DWT-based compression enables the scalability of images since at each stage of decomposition it allows them to be split into four subbands. A hybrid watermarking scheme can be developed by selecting the appropriate subband to increase robustness, imperceptibility, and capacity [8].

2 Literature Survey

Numerous techniques have been proposed to hide watermarks in digital images. Various criteria allow further classification of these techniques into various categories. Spatial and transform domains are the two main domain types. The application-specific needs and limitations determine which domain, i.e., frequency or spatial, should be used to include a watermark. Better resistance to image processing operations such as compression, filtering, and rotation is offered by the frequency domain [9]. This is because the watermark can be incorporated into the modified image's higher frequency coefficients, which are less likely to be impacted by these procedures. On the other hand, with spatial domain watermarking, it is possible within the pixel values of the image to embed the watermark directly. This increases the watermark's susceptibility to changes in the pixel value during image processing.

2.1 Spatial Domain

LSB-based Watermarking Watermarks are embedded via the Least Significant Bits of pixel values in the cover image as shown in Fig. 3. It is the simplest watermarking technique. The disadvantages of LSB replacement include unwanted noise, cropping, lossy compression, etc. This method encrypts the message image with the cover image using watermarking and the LSB technique, then decrypts the message image and cover image to restore them to their original state. To implement image security, an effective spatial domain approach LSB is employed. The most effective method of image watermarking is determined using this method, where the significant bit ranges from 1 to 8 [10].

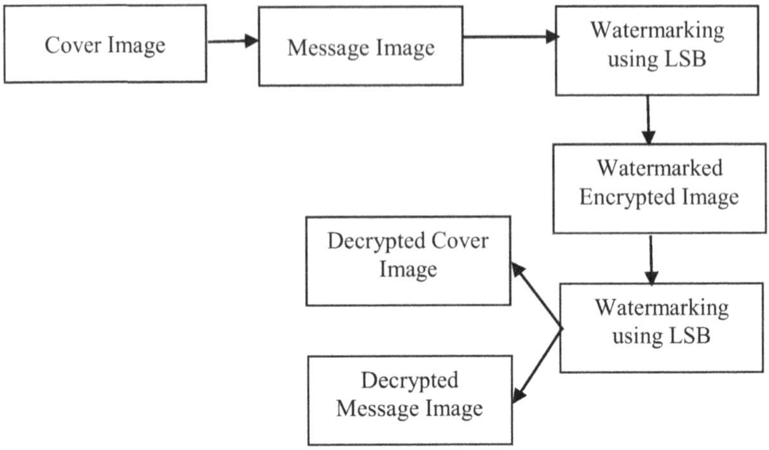

Fig. 3 Watermark Embedding and Extraction Process

Patchwork Technique Patchwork technique is one of the statistical techniques [11]. Using a Gaussian distribution as a basis, watermarks are inserted into the image through statistical analysis as patches. The working principle of this technique is explained as follows: Two of the patches are arbitrarily chosen, say patch A and patch B. The brightness of Patch A's image-data is increased, while that of Patch B is decreased, resulting in contrasting levels of brightness between the two patches. In this method, redundant pattern encoding is used to insert information into an image [11].

2.2 *Frequency Domain*

Discrete Wavelet Transform (DWT) DWT has a diverse range of applications in signal processing tasks, such as simulating wireless antenna dispersion, reducing noise in audio and video, and compressing audio and video data. The method described involves the conversion of an RGB image to YCbCr and watermarking using DWT [12]. For embedding watermark, the luminance component Y of the image is reviewed. The proposed technique is assessed for red, blue, and green using metrics such as PSNR, SNR, MSE, and NC. The results of this chapter have shown that the technique is very effective for watermarking authentication and promotes improved security and precise alignment of the extracted watermark with the original watermark [13].

Discrete Cosine Transform (DCT) A common method of transform domain watermarking is a DCT. It allows for the selection of an appropriate frequency range for watermark embedding by dividing the image into components of high, middle, and low frequency. Since the middle-frequency bands prevent the spread of watermark information to the majority of visually significant areas of the image, they are typically used in watermarking. The proposed technique involves the combination of DCT and DWT to embed and extract the watermark. This approach is used to ensure that the target images can be made available on the Internet with the least possible compression while maintaining the robustness of the watermark. In other words, the technique is designed to make sure that the watermark can survive the compression process and still be identifiable, even if the file is intentionally compressed to make it smaller for uploading on the Internet [14].

Singular Value Decomposition (SVD) SVD uses one-way non-symmetrical decomposition and non-fixed orthogonal bases instead of fixed orthogonal bases like those found in DCT, DFT, and DWT transformations; it differs significantly from those transformations. SVD provides the benefits of different transformation sizes and increased security. It can perform well in terms of security and robustness [15].

3 The Proposed Zero Watermarking Methodology for Data Authentication

In this work, a data authentication technique based on DCT and DWT is demonstrated on images in the frequency domain. The DCT is a mathematical technique that is used to analyze digital signals and data, particularly in image and video processing applications. It entails transforming a signal with its frequency components from the spatial domain, which contains the pixel values in an image, to the frequency domain. By altering the frequency components of the image, DCT is frequently employed in digital image watermarking to insert a watermark into the image. A unique watermark can be embedded in a picture by adjusting certain DCT coefficients. The watermark inserted in the image can be removed later. The watermark can then be used to determine the image's owner or origin. One benefit of employing DCT for image watermarking is that it is a procedure that can be reversed; thus, even after the watermark has been inserted, the original image can be restored with no loss of quality. DCT is also resilient to standard image processing procedures including resizing and compression, which makes it a reliable method for watermarking digital photos [16]. The original image is divided into 8 by 8 pixel blocks for the DCT to work on, and each block is then transformed independently. The middle-frequency range of the DCT coefficients is extracted using a 2D mask. This mask picks out particular DCT coefficients in the block, which can subsequently be changed to include a watermark. The watermark can withstand attacks such as compression and cropping better with this technique. A one-dimensional DCT can be applied to the analysis of one-dimensional signals. However, a two-dimensional version of the DCT is required for the analysis of a two-dimensional signals, such as pictures. The mathematical formulas for the 2D DCT and 2D

Inverse DCT (IDCT) transform are represented by Eqs. (1) and (2), respectively. The updated DCT coefficients are encoded and decoded using these transforms to retrieve the watermark from the watermarked image.

$$F(m,n) = \sum_{i=0}^{N-1} \sum_{j=0}^{N-1} c(m)c(n)f(i,j)\text{coscos}\left[\frac{\pi(2i+1)m}{2N}\right] * \cos\left[\frac{\pi(2j+1)m}{2N}\right] \quad (1)$$

$$f(i,j) = \sum_{i=0}^{N-1} \sum_{j=0}^{N-1} c(m)c(n)F(m,n)\cos\left[\frac{\pi(2i+1)m}{2N}\right] * \cos\left[\frac{\pi(2j+1)m}{2N}\right] \quad (2)$$

where

$$c(m), c(n) = \{\sqrt{\frac{1}{N}} \sqrt{\frac{2}{N}} \quad \begin{matrix} m,\ n = 0 \\ m,n = 1\,\text{up to}\,N-1 \end{matrix}$$

In watermarking, the DWT breaks down a signal into various frequency subbands that can be used to represent the signal in a more manageable and reliable manner. The DWT is frequently employed in the context of watermarking in order to embed the watermark into the subbands that are least apparent to human perception while preserving the original signal's quality. The watermark can be added to the DWT subband coefficients by either multiplying or adding a watermark signal to the subband coefficients. The watermarked signal can be retrieved by reconstructing the signal using the Inverse DWT (IDWT). The watermark can be retrieved by calculating the DWT of the watermarked signal and contrasting the subband coefficients with those of the original signal [17]. Because DWT coefficients capture both the local and global properties of the signal, they are resistant to various attacks including compression, noise, and geometric alterations. However, attacks such as cropping, resampling, or filtering can change the subband coefficients and make it challenging to extract the watermark. DWT is a technique that can be employed to reduce the dimensionality of an image and perform feature extraction. DWT algorithm decomposes the image into four subbands, i.e., LL, LH, HL, HH [18].

- LL is the approximate image of the input image; it is a low-frequency subband so it is used for the further decomposition process.
- LH subband extracts the horizontal features of the original image.
- HL subband gives vertical features.
- HH subband gives diagonal features.

An image can be processed with flexibility and efficiency by breaking it down into its component bands, LL, LH, HH, and HL. This multi-resolution representation can be used for a variety of image-processing applications.

Hamming Distance The Hamming distance is a metric used to compare two binary data strings. When two binary strings of the same length are compared, the Hamming distance is the count of the number of bit positions in which the two bits differ. The Hamming distance between two strings, *a* and *b*, is denoted by *d(a, b)*. Error detection and correction techniques are often employed, when transmitting data over computer networks, to ensure the accuracy of the transmitted data. It is also used in coding theory to compare data words of the same length [19].

The images that are stored in the directories should be fetched directly using *cv2.imread( )* which is present in the OpenCV module. The image is divided into blocks of size 8 × 8 through a nested for loop which has a range from 0 to height-8 and from 0 to breadth-8 which jumps in a range of 8. The image is sliced from the *index* to *index+8* for further processing as shown in Fig. 4. The DCT is applied on one block of size 8x8 by using *dct( )* function from *scippy.fftpack* library and setting *norm='ortho'*. Further, DWT is applied on the same block to get the subbands LL, LH, HL, HH. The mean value is calculated for the LL approximation image using the mean( ) function and is appended in a list. Entropy using *skimage.measure.shannon.entropy( )* for the same LL approximation image is calculated and appended in a separate list. For the mean value, the list compares the adjacent elements, and if the first element is greater than or equal to the second element, then '1' is appended to a string or else '0' is appended to the same string. The entropy values are computed for the list created. XOR operation is performed on mean value string and entropy string on corresponding positions and the key value is stored. Various attacks such as Gaussian, salt, and pepper, mean and median are applied on the image separately. The above-mentioned steps are repeated till the specified threshold is reached and are illustrated in Fig. 5. The analysis is performed by calculating the variations in the number of bits and Hamming distance.

4 Experimental Results

The accuracy and reliability of the proposed watermarking method are evaluated by testing it on various host and watermark images. The Hamming distances computed are shown in Table 1. The Hamming distance is a metric that measures the number of bit positions in which two code words differ [20]. The performance of the proposed methodology is analyzed by introducing various noise to the images. Table 1 provides the results of a study analyzing the effects of different types of attacks on images with various message sizes. Here the bit error is the number of erroneous bits introduced during the attack. It indicates the level of distortion or corruption in the image due to the attack. The Hamming distance measures the difference between two strings of equal length. In this context, it quantifies the dissimilarity between the original image and the distorted image after the attack. It is calculated as the number of positions at which the corresponding bits are different. This table illustrates how different attacks affect images with varying message sizes. Lower values of bit error and Hamming distance suggest that the attack has caused less distortion in the image, indicating better robustness against the attack. Conversely, higher values indicate more significant distortion and reduced robustness.

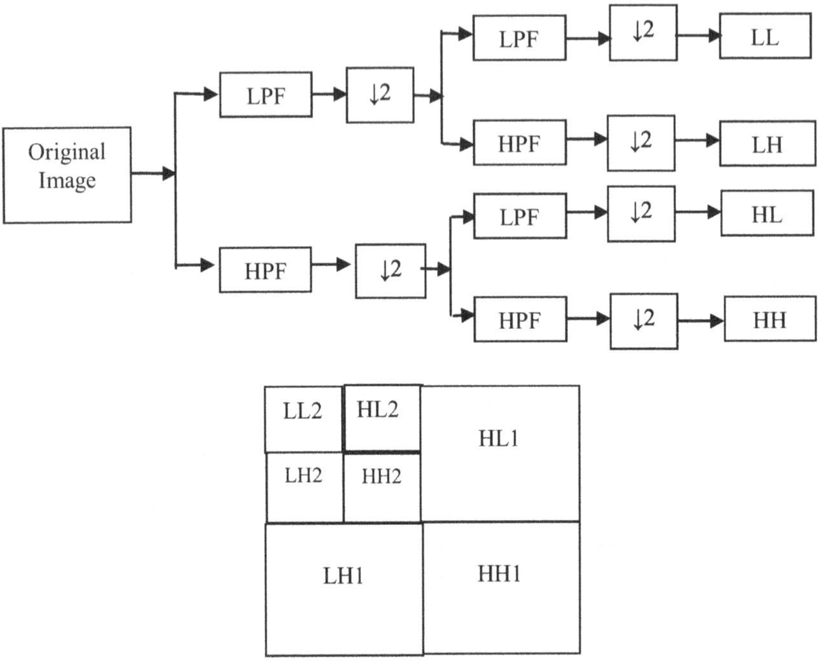

Fig. 4 Two-level DWT Decomposition Process

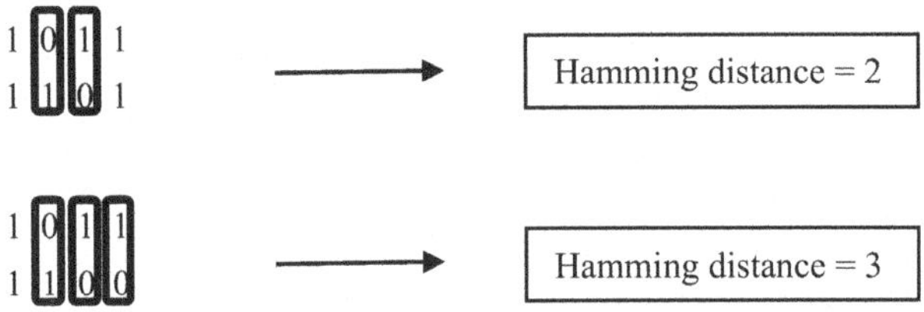

Fig. 5 Computation of Hamming Distance

Gaussian Gaussian noise is a common type of attack in image processing and can have negative effects on the quality of the image as shown in Fig. 6b. The noise can make the image appear grainy or blurry and can reduce the overall visual quality and information content of the image. Gaussian noise can be added to an image in different ways, such as by using random values generated from a Gaussian distribution with zero mean and a certain standard deviation, or by applying a Gaussian filter to the image. The effect of this noise on the image depends on the standard deviation of the Gaussian distribution used. Higher values of the standard deviation will result in more noise and more severe degradation of the image.

Salt and Pepper Attack Salt and pepper noise is a common type of attack on digital images, caused by randomly occurring bright and dark pixels as shown in Fig. 6c.

Table 1 Computation of Hamming Distance for Various Attacks

Attacks	Message Size (in bits)	Bit Error	Hamming Distance
Gaussian	48	0	0.00
Mean	48	0	0.00
Median	48	8	0.17
Salt and pepper	48	15	0.31
Gaussian	55	8	0.15
Mean	55	1	0.02
Median	55	1	0.02
Salt and pepper	55	1	0.02
Gaussian	63	8	0.13
Mean	63	0	0.00
Median	63	8	0.13
Salt and pepper	63	16	0.25
Gaussian	120	12	0.10
Mean	120	12	0.10
Median	120	12	0.10
Salt and pepper	120	11	0.09
Gaussian	168	1	0.01
Mean	168	15	0.09
Median	168	14	0.08
Salt and pepper	168	43	0.26

The bright pixels are known as *salt* while the dark pixels are known as *pepper*. The salt and pepper noise can corrupt the image data, resulting in loss of detail and distortion of image features.

Median Filtering Attack Median filtering is used in image processing to reduce the noise present in an image and improve its quality as shown in Fig. 6d. The most widely used non-linear filter, known as the Median Filter, employs order statistics to rank the pixels present in the image area covered by the filter and then replaces the value center pixel with the median of the intensity values nearby. Median filters are particularly effective at reducing certain types of random noise and produce less blurring than linear smoothing filters of a similar size, making them a popular choice. In this experiment, a 5 × 5 pixel Median Filter window size is applied to the watermarked images produced. The attacked watermarked images are then processed to extract the watermark signal, which is compared to the original signal.

Mean Filtering The mean attack works by replacing each pixel in the image with the average value of its neighboring pixels. The size of the neighborhood can vary depending on the specific implementation of the attack. A larger neighborhood size will result in more blurring and a greater loss of detail in the image as shown in Fig. 6e. This type of attack can be used to blur an image and reduce its sharpness.

Fig. 6 (**a**) Original Image (**b**) Gaussian Attack (**c**) Salt and Pepper Attack (**d**) Median Filtering Attack (**e**) Mean Filtering Attack

The proposed work addresses the above-mentioned issues with security, authentication, by safeguarding intellectual property. The future of digital content protection and management in the areas of digital multimedia and watermarking techniques may be shaped by the fusion of cutting-edge watermarking techniques with developing technologies.

5 Conclusions

The proposed Discrete Cosine Transform (DCT) and Discrete Wavelet Transform (DWT) watermarking algorithm provides better defense mechanisms against several attacks. This assertion is substantiated by the experimental results where the performance of the proposed algorithm is evaluated against various adversarial scenarios, including mean, median, salt and pepper, and Gaussian attacks. In the proposed watermarking methodology, characteristics such as low-frequency band for entropy calculation and mean value determination in the image, are strategically leveraged. This deliberate selection empowers the proposed algorithm with the ability to extract feature information from regions where wavelet coefficients are more stable, thus enhancing the resilience of the zero-watermark. Additionally, the incorporation of the XOR operation bolsters the security posture of the proposed approach, mitigating potential vulnerabilities against malicious tampering. The use of Hamming distance plays a significant role in the assessment of algorithmic efficacy. This metric encapsulates the degree of dissimilarity between the original and distorted images, thereby providing a quantitative measure of the algorithm's robustness under attack scenarios. The experimental results consistently demonstrate that the proposed algorithm maintains higher image quality and robustness across varying message sizes, as presented in Sect. 4. Specifically, our findings reveal that the proposed algorithm yields lower bit error rates and low Hamming distances across different message sizes when compared to conventional watermarking techniques. Notably, even under the most rigorous Gaussian attack, the proposed algorithm exhibits commendable resilience, with minimal degradation in image quality and robustness. These compelling results underscore the efficacy and practical viability of the proposed DCT and DWT based watermarking algorithms. By offering a judicious balance between security, robustness, and image quality preservation, the proposed approach presents a promising solution for safeguarding digital media assets against a myriad of potential threats and attacks. The proposed algorithm demonstrates better performance in handling Gaussian and Median attacks compared to Mean and Salt and Pepper attacks, particularly for message sizes of 128-bit.

References

1. Mahbuba Begum and Mohammad Shorif Uddin. "Digital image watermarking techniques: a review". Information, Vol. 11(2), No.110, pp.1–38, 2020.
2. Md. Maklachur Rahman, "A DWT, DCT, and SVD-based watermarking technique to protect the image piracy", International Journal of Managing Public Sector Information and Communication Technologies, Vol. 4, No. 2, pp.21–32, 2013.
3. Rajesh Kannan Megalingam, Mithun Muralidharan Nair, Rahul Srikumar, Venkat Krishnan Balasubramanian, and Vineeth Sarma Venugopala Sarma. "Performance comparison of novel, robust spatial domain digital image watermarking with the conventional frequency domain watermarking techniques". In IEEE International Conference on Signal Acquisition and Processing, pp. 349–353, 2010.
4. Thottempudi Pardhu and Bhaskara Rao Perli. "Digital image water- marking in frequency domain". In IEEE International Conference on Communication and Signal Processing (ICCSP), pp. 208–211, 2016.
5. Bhupendra Ram. "Digital image watermarking technique using discrete wavelet transform and discrete cosine transform", International Journal of Science and Research (IJSR), Vol. 10, No. 12, 2021.
6. Preeti Gaur and Neeraj Manglani. "Image watermarking using LSB technique". International Journal of Engineering Research and General Science, Vol. 3, No.3, pp.1424–1433, 2015.
7. Hiroyuki Kii, Junji Onishi, and Shinji Ozawa. "The digital watermarking method by using both patchwork and DCT". In Proceedings IEEE International Conference on Multimedia Computing and Systems, Vol. 1, pp. 895–899, 1999.
8. Pravin M Pithiya and HL Desai. "DWT based digital image watermarking, de-watermarking & authentication. Image", Vol. 7, No. 5, 2013.
9. Atheer Bassel and Md Jan Nordin. "Digital image watermark authentication using DWT-DCT",. Journal of Engineering and Applied Sciences, Vol.11, No.14, pp.3227–3232, 2016.
10. Tribhuwan Kumar Tewari and Vikas Saxena. "An improved and robust DCT based digital image watermarking scheme", International Journal of Computer Applications, Vol.3, Issue No.1, pp.28–32, 2010.
11. Chin-Chen Chang, Piyu Tsai, and Chia-Chen Lin. "SVD-based digital image watermarking scheme. Pattern Recognition Letters, Vol. 26, No..10, pp.1577– 1586, 2005.
12. Ali Al-Haj, "Combined DWT-DCT digital image watermarking", Journal of computer science, Vol.3, No..9, pp.740–746, 2007.
13. BP Rinky, Payal Mondal, K Manikantan, and S Ramachandran. "DWT based feature extraction using edge tracked scale normalization for enhanced face recognition", Procedia Technology, Vol.6, pp.344–353, 2012.
14. Prabhjot Kour., "Image processing using discrete wavelet transform", International journal of electronics and communication, Vol.3, No.1, 2015.
15. Mohammad Norouzi, David J Fleet, and Russ R Salakhutdinov. "Hamming distance metric learning", Advances in neural information processing systems, Vol. 25, pp.1–9, 2012.
16. Chunlin Song, Sud Sudirman, Madjid Merabti, and David Llewellyn- Jones. "Analysis of digital image watermark attacks", In 7th IEEE Consumer Communications and Networking Conference, pp. 1–5, 2010.
17. K. Hameed, M. S. Khan, I. Ahmed, Z. U. Ahmad, A. Khan, A. Haider, N. Javaid, "A Zero Watermarking Scheme for Data Integrity in Wireless Sensor Networks," 19th International Conference on Network-Based Information Systems (NBiS), Ostrava, pp. 119–126, 2016.
18. H. Huang, L. Zhang and Z. Wu, "Secure and Reliable Multidimensional Orthogonal Code aided RF Watermark Design for NB-IoT Systems," IEEE International Conference on Communications, Ireland, pp. 1–6, 2020.
19. E. Elbasi, "B-DCT based Watermarking Algorithm for Patient Data Protection in IoMT," International Conference on Information Security and Cryptology (ISCTURKEY), Ankara, Turkey, pp. 1–4, 2020.
20. Y. Xiao and G. Gao, "Digital Watermark-Based Independent Individual Certification Scheme in WSNs," in IEEE Access, Vol. 7, pp. 145516–145523, 2019.

Gender-Based Differences in Demographic and Time Domain Pulse Rate Variability Parameters in Association with Type2 Diabetic

Shikha Agarwal and Rakesh Kumar Sinha

1 Introduction

Diabetes is a metabolic disorder where the blood glucose level remains high than certain normal value for a longer period of time. The most common type of diabetes is type 2 diabetes, which is a progressive illness mainly due to high insulin resistance and beta cell dysfunction because of seductive life style, obesity, and stress [1]. The IDF Atlas (2021) reported [2] that approximately 537 million adults between 20 and 79 years are living with diabetes, which is 10.5% of world population. More than 90% of world population has T2DM with key contributions as urbanization, aging, decreased physical activity, and obesity [3]. According to IIPS, 3 out of 4 diabetic adults live in low- and middle-income countries [4]. IDF Atlas [2] says that prevalence of diabetes in males (10.8%) is slightly higher than females (10.2%). Similar study from the ICMR-INDIA reported that occurrence of diabetes in males in middle age group (35–65 years) is slightly higher in females [5].

There is increased evidence that cardiovascular disease and type 2 diabetes share common metabolic risk factors [6–9]. These studies show that prevalent risk factors are age, smoking, obesity, hypertension, and dyslipidemia, which leads to polyneuropathy, retinopathy, nephropathy, erectile dysfunction, etc. Many gender-based

S. Agarwal (✉) · R. K. Sinha
Department of Bioengineering and Biotechnology, Birla Institute of Technology, Mesra, Ranchi, India
e-mail: phdbe10051.17@bitmesra.ac.in; rakeshsinha@bitmesra.ac.in

A. Haldorai et al. (eds.), *6th EAI International Conference on Big Data Innovation for Sustainable Cognitive Computing*, EAI/Springer Innovations in Communication and Computing, https://doi.org/10.1007/978-3-031-54696-9_7

studies have also been done to evaluate the severity of autonomic impairment between male and female. A random study [10] on prevention of diabetes reported that men are at higher risk of diabetes. Another study [11] reported that women have more weight-related psychosocial problems. Similarly, in a study [12] for association of gender-related characteristics and type 2 diabetes risk in a multi-ethnic population concluded that the results were consistent between male and female, and there is no formal evidence for significant gender variables with T2DM. In a review article [13] reported that diabetic women have 40% more risk of CHD compared with male patients [14, 15], the higher progression of DKD in men [16–18], and there is higher comorbidities in women [19, 20]. Most of the above studies were based on clinical results both in terms of current and follow-up measurements done on selected subjects based on the objective of the study. To overcome the difficulty in patient management and longer study periods, some researchers have also done sex-related study in association of diabetes based on HRV parameters to study in non-invasive way from ECG recordings [21–27]. The health professionals are required to understand and consider the different health needs and pathogenesis if exists, between women and men. HRV has been widely used for cardiac and neuron autonomic activities assessment because of its direct relationship between cardiac rhythm and activity of parasympathetic-sympathetic nervous system through ECG signals. But due to difficulty in recording good quality ECG signals, another technique, namely, PRV has been used in recent years for HRV analysis using pulsed waveform mainly from PPG technique [28–32]. Because of its simplicity and widespread use of PPG [33], this study is an approach to study gender-based differences in type2 diabetic status both from demographic and PRV analysis through finger PPG recordings. Non-diabetic group were also studied to find out any difference in parametric values for males and females. This study can be proved an important and a novel approach for many reasons. FPPG data recording is very safe, easy, and economical. No special patient preparation or training is required for PPG data recordings. Patient from any age group (mostly) can be taken for the study in a relaxed sitting position, which gives stable data recordings over ECG.

2 Methods

2.1 Subjects

The FPPG data is recorded both for diabetic and control group subjects. A total number of 486 subjects were participated where 252 subjects were diagnosed with diabetes having HbA1c $\geq$ 6.5% (as per WHO report 2011 on diabetes [34]) and 234 were considered under control group. Out of 486 participants, 284 were males and 202 were females (Table 1).

Table 1 Gender-based demographic and clinical characteristics of the study participants

	Male ($N = 284$) Mean ± SD	Female ($N = 202$) Mean ± SD	Total ($N = 486$)	p-value (two-tail, $p < 0.05$)
Age	50.52 ± 13.92	41.78 ± 16.51		5.17E−05
Weight	71.26 ± 13.7	60.64 ± 10.87		1.76E−09
Height in cm	165.64 ± 15.8	155.08 ± 6.45		8.87E−10
BMI	25.58 ± 4.8	25.23 ± 4.37		0.58717
HR	85.95 ± 16.21	89.22 ± 10.94		0.082956
Systolic blood pressure	128.5 ± 14.39	124.50 ± 17.09		0.068864
Diastolic blood pressure	82.04 ± 10.36	77.98 ± 8.7		0.002122
Diabetic	150	102	252	
Non-diabetic	134	100	234	

2.2 *Recording Protocols*

The data were recorded according to the Ethical Approval protocol prescribed by Ranchi Urology Centre, Bariyatu Road, Ranchi. The volunteers filled out a standard detailed questionnaire on their blood and clinical reports, demographic and personal data, past medical and family history, habits, and medications. The consent for data recordings was also taken in the language understood by the participant prior to recording. The participants were asked to sit in the relaxed position for 5–10 minutes, and then recording of PPG signals were done from their left hand index finger at normal room temperature in stable condition. The inclusion criteria for study groups were age ≥18 years and age ≤80 years, systolic blood pressure ≤220 mmHg, and with and without CHD. Exclusion criteria were any kind of neuropathy, transgender subject, recent hospitalization for past 10 days, pregnant or unstable subject, and type1 diabetes.

2.3 *PRV Analysis*

PRV is the change in pulse rate caused by changes in ANS activities due to different external environmental and internal hormonal changes in human body [35]. This study used finger PPG to get pulsed waveforms for the PRV measurements and analysis. The 5-minute PPG recording was done with the help of portable PLUX Wireless Bio signals device S.A., from Portugal for each participant from both study groups. The data are recorded at the sampling rate of 200 sampling points per second. Since the device is very robust with inbuilt filter, the PPG data were recorded without disturbance and saved in different file format for offline analysis. No further filter was applied during data preprocessing phase, and all analysis was done on raw data only to ensure that no information should be lost. The saved PPG data in .txt file is postprocessed, and PRV analysis was done through BIOPAC STUDENT

LAB 4.0 software. The sample rate interval was set as 200 Hz, and the PPG data were plotted as PPG waveforms. The raw tachogram was obtained after baseline correction. Any unwanted spikes and turfs were edited and again 5-minute recorded wave data was copied and saved in .txt file for further PRV measurements and analysis. KUBIOS HRV standard 3.5.0 version was used to calculate time domain PRV measurements and report generation for further analysis. The time domain PRV parameters measured are listed in Table 2. PRV reflects variations between

Table 2 Summary of time domain PRV measures calculated in this study

Measures	Description
Mean RR*(ms)	Mean value of all RR intervals (longer RR interval means lower HR and higher parasympathetic activity) [36]
Mean HR*(bpm)	Mean value of heart rates. HR is calculated as $$\mathrm{HR} = \frac{60}{\mathrm{RR\,in\,seconds}} \quad (1)$$ (higher mean HR indicates sympathetic dominance) [36]
Min HR(bpm)	The lowest HR value during entire data recording
Max HR(bpm)	The highest HR value during entire data recording
RMSSD(ms)	Square root of mean of squares of successive RR interval differences (higher value indicates high parasympathetic activity) [36] $$\mathrm{RMSSD} = \sqrt{\frac{1}{n-1}\sum_{k=1}^{n-1}\left(\mathrm{RR}_{k+1} - \mathrm{RR}_k\right)^2} \quad (2)$$ where n = total no. of observations
SDNN(ms)	It is standard deviation of all RR intervals given as [36] $$\mathrm{SDNN} = \sqrt{\frac{1}{n-1}\sum_{k=1}^{n-1}\left(\mathrm{RR}_k - \overline{\mathrm{RR}}\right)^2} \quad (3)$$ where $\overline{\mathrm{RR}}$ is mean of all RR intervals
NN50(beats)	No. of pairs of consecutive RR intervals that differ by more than 50 milliseconds in whole recording (higher value indicates parasympathetic activity dominance) [36]
pNN50 (%)	$$\mathrm{pNN50} = \frac{\mathrm{NN50}}{\mathrm{Total\,no.of\,RR\,intervals}} * 100\% \quad (4)$$
RR Triangular index	Integral of the density of RR interval histogram divided by the height of the histogram [36, 37]
TINN(ms)	It is the baseline width of RR interval histogram [36, 37]
Stress index	It is a geometric measure of PRV given by square root of Baevsky's stress index [38]. Higher value indicates reduced variability and increased sympathetic activity

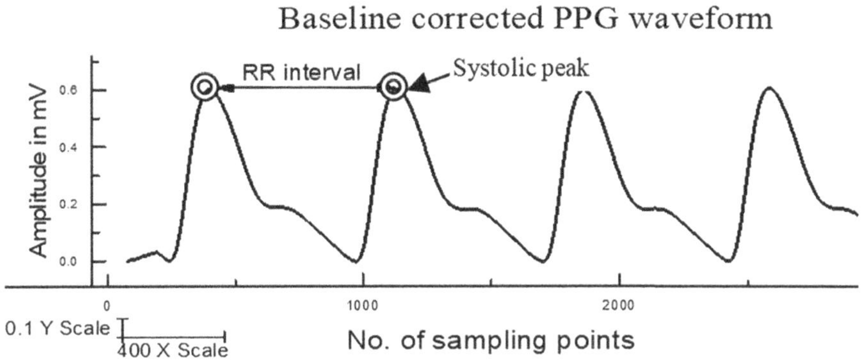

Fig. 1 Finger PPG waveforms after baseline correction with peak-to-peak RR interval

consecutive RR intervals. RR interval is measured as time difference between two consecutive systolic peaks of PPG waveforms as shown in Fig. 1. Time domain PRV measures are the values that can be obtained from statistical analysis of the time intervals between the heart beats. It can be successive or interval heart beats from 24-hour data recordings to few minutes' recordings.

2.4 *Statistical Analysis*

The measured PRV parameters and relationship graphs were generated as a complete report for each participant through Kubios software. All PRV parametric values and demographic data of each participant are stored in Excel sheet for their statistical analysis. The statistical values such as mean, SD, SE, and variance were calculated for each variable. The clinical and demographic characteristics of studied participants were assessed for male and female by analysis of variance (continuous variable) (Table 1). The significance test was done using two-tailed Student's t-test method. The significance level was set at 5%. Also, correlation coefficient "*r*" is calculated for each PRV parameters and demographic data with respect to diabetes through Pearson's correlation analysis method both for males and females (Table 3). The independent association between HRV parameters and demographic data was evaluated using multivariable linear regression) both for male and female. Differences in values between different categorical groups such as diabetic-non-diabetic, male-female in all combinations were tested with Student's paired two-tailed t-test where $p < 0.05$ was considered as significant. Values are reported as mean ± SD (Table 4).

Table 3 Pearson's correlation coefficient between diabetes mellitus and variables for male and female

Parameters	Male (*N* = 112) Correlation coefficient "*r*" w.r.t diabetes	Female (*N* = 99) Correlation coefficient "*r*" w.r.t diabetes
Age (years)	0.438	0.727
Weight (Kg.)	0.013	0.209
Height (cms.)	−0.004	−0.218
BMI	0.050	0.312
HR	0.353	−0.169
SBP(mmHg)	**0.226**	**0.567**
DBP(mmHg)	**0.112**	**0.486**
Stress index	**−0.007**	**0.288**
Mean RR*(ms)	−0.205	0.132
Mean HR*(bpm)	0.207	−0.117
Min HR (bpm)	0.015	−0.027
Max HR (bpm)	0.160	−0.255
SDNN (ms)	0.114	−0.003
RMSSD (ms)	**0.165**	**0.027**
NN50 (beats)	0.003	−0.075
pNN50 (%)	−0.056	−0.238
RR triangular index	−0.105	−0.075
TINN (ms)	0.133	0.058

3 Results

The finger PPG data were recorded both for male and female subjects to study their correlation and differences in association with diabetes mellitus and also for healthy condition. To evaluate significance of correlation or differences, the variables were demographic data of the subjects such as age, weight, blood pressure (Table 1), and the PRV measures (Table 2).

3.1 *Demographic Gender-Based Differences in Overall Studied Population*

It can be observed from Table 1, which is a summary of demographic and clinical characteristics of study participants, that there is significant difference with *p*-value 0.002122 ($p < 0.05$) for diastolic blood pressure between male and female for overall population. This may be due to the average age of studied female population of females is lower (41.78 ± 16.51) than males (50.52 ± 13.92). There is no significant difference observed in HR and BMI for overall studied population, which is speculated also.

Table 4 Significance test between all combination groups for diabetes mellitus and gender for calculated variables

Variable	Diabetic			Non-diabetic			Female			Male		
	Female	Male		Female	Male		NDM	DM		NDM	DM	
	MEAN ± SD	MEAN ± SD	*p*-value	MEAN ± SD	MEAN ± SD	*p*-value	MEAN ± SD	MEAN ± SD	*p*-value	MEAN ± SD	MEAN ± SD	*p*-value
Age (years)	31.0 ± 11	42. ± 13	**2.28E−05**	55.8 ± 11	53.7 ± 14	0.3054	55.5 ± 11	31.0 ± 11	**2.16E−17**	53.7 ± 14	42.7 ± 13	**4.15E−05**
Weight (Kg.)	58.5 ± 12	71.0 ± 13	**2.79E−06**	63.2 ± 9	68.0 ± 13	**0.0015**	63.2 ± 9	58.6 ± 12	**0.0224**	68.0 ± 13	71.0 ± 13	0.2248
Height In cm	156.3 ± 6	162.3 ± 9	0.0587	153.4 ± 5	162 ± 24	**8.75E−15**	153.4 ± 5	156.3 ± 6	**0.0167**	162. ± 24	162 ± 9	0.8779
BMI	24 ± 4.5	25.8 ± 4	**0.0292**	26.7 ± 3	25.6 ± 5	0.1587	26.79 ± 3	24 ± 4	**0.0010**	25.62 ± 5	25.88 ± 4	0.7666
HR (bpm)	91.7 ± 10	78.2 ± 15	**1.18E−07**	87.1 ± 10	88.2 ± 10	0.1110	87.1 ± 10	91 ± 10	**0.0473**	88.2 ± 10	78.6 ± 15	**9.64E−06**
Systolic pressure (mmHg)	115.4 ± 19	124.6 ± 13	**0.0003**	135.2 ± 8	129.7 ± 13	0.0752	135.5 ± 8	115.9 ± 19	**3.77E−08**	129.7 ± 13	124.3 ± 13	**0.0288**
Diastolic pressure (mmHg)	74.18 ± 7	80.55 ± 8	**0.0015**	82.79 ± 7	81.9 ± 11	0.4697	82.7 ± 7	74.1 ± 7	**1.50E−07**	81.9 ± 11	80.5 ± 8	0.4731
Mean RR* (ms)	746 ± 263	743 ± 124	0.4704	745 ± 118	758 ± 140	0.3057	745 ± 118	746 ± 263	0.4901	758 ± 140	743 ± 124	0.2658
Max HR (bpm)	100 ± 19	99.4 ± 20	0.3921	98.4 ± 24	103 ± 27	0.1357	98.4 ± 24	100. ± 19	0.3258	103 ± 27	99.4 ± 20	0.4566
SDNN (ms)	79 ± 151	99. ± 135	0.2301	76 ± 156	83.4 ± 78	0.37564	76 ± 156	79 ± 151	0.45414	83.4 ± 78.17	99.18 ± 135.72	0.181114891
RMSSD (ms)	110.7 ± 226.12	136.66 ± 188.01	0.252854555	93.08 ± 199.21	102.25 ± 99.07	0.378992938	93.08 ± 199.21	110.7 ± 226.12	0.340872012	102.25 ± 99.07	136.66 ± 188.01	0.073186309
NN50 (beats)	45.26 ± 78.39	49.36 ± 53.96	0.376116207	42.54 ± 41.67	62.59 ± 50.05	**0.00989957***	42.54 ± 41.67	45.26 ± 78.39	0.416660398	62.59 ± 50.05	49.36 ± 53.96	0.064024495

(continued)

Table 4 (continued)

Variable	Diabetic			Non-diabetic			Female			Male		
	Female	Male		Female	Male		NDM	DM		NDM	DM	
	MEAN ± SD	MEAN ± SD	*p*-value	MEAN ± SD	MEAN ± SD	*p*-value	MEAN ± SD	MEAN ± SD	*p*-value	MEAN ± SD	MEAN ± SD	*p*-value
pNN50 (%)	12.2 ± 15.6	17.49 ± 21.02	0.050667818	18.88 ± 16.44	23.22 ± 20.69	0.104912479	18.88 ± 16.44	12.2 ± 15.6	**0.019640***	23.22 ± 20.69	17.49 ± 21.02	**0.049391***
RR triangular index	7.78 ± 11.63	8.15 ± 5.23	0.421115534	8.64 ± 3.6	9.79 ± 4.68	0.068098478	8.64 ± 3.6	7.78 ± 11.63	0.316589479	9.79 ± 4.68	8.15 ± 5.23	**0.0366827***
TINN (ms)	570.65 ± 962.3	664.55 ± 909.42	0.289622072	428.16 ± 744.96	529.3 ± 541.12	0.203997064	428.16 ± 744.96	570.65 ± 962.3	0.207372706	529.3 ± 541.12	664.55 ± 909.42	0.11919662
Stress index (SI)	12.81 ± 6.74	10.98 ± 7.19	0.078990885	11.28 ± 5.94	9.14 ± 4.72	**0.01747852***	11.28 ± 5.94	12.81 ± 6.74	0.116535087	9.14 ± 4.72	10.98 ± 7.19	**0.0351797***

3.2 Gender-Based Relationship Between Variables and DM

The gender-based relationship in association with DM is well summarized in Table 3 for all measured variables. Most of the variables do not show good correlation with DM. But it is strange to observe that some variables showed opposite correlation with DM for males and females. The Pearson's correlation coefficient "*r*" shows positive correlation ($r = 0.288$) with DM for females, which was expected but minor negative correlation ($r = -0.007$) with DM for males. However, the correction is highly insignificant, but negative correlation is strange to notice. Similarly, mean RR ($r = -0.205$) is negatively correlated for males and positive ($r = 0.132$) for females. The same relation is confirmed by HR where it is positive ($r = 0.207$) for males and negative ($r = -0.117$) for females as RR and HR are inversely proportional to each other. The same is true for minimum and maximum HR also. Standard deviation of RR intervals (SDNN) also showed positive correlation for males ($r = 0.004$) and negative for females (−0.003) with respect to DM. Though the correlation is insignificant but opposite correlation for men and women is not speculated and the reason for this observation cannot be explained also in the capacity of this study.

It is obvious to observe that both male and females are positively correlated for age vs. DM (highest correlation), but females showed higher correlation ($r = 0.727$) than males ($r = 0.438$). The conclusion can be here that the risk to have diabetes with age in females is higher than males. Similarly, both systolic and diastolic blood pressures along with weight are positively correlated with DM, which shows that patients with obesity and diabetes have higher blood pressures. Like aging factor females tend to have higher correlation for systolic ($r = 0.567$) and diastolic ($r = 0.486$) blood pressure, weight ($r = 0.209$), BMI ($r = 0.312$) with respect to DM than males ($r = 0.226$ (SBP), 0.112 (DBP), 0.013 (Wt.), 0.050 (BMI)). So obese and hypertensive females are at higher risk to have diabetes than to males.

Bar and distribution graph is plotted for males and females between demographic variables of the studied participants and DM to compare categorically in Fig. 2. It can easily be seen from Fig. 2a that diabetic females are in higher age group (50–60 years) than non-diabetic females (25–35 years) whereas diabetic males also have higher age(40–60 years) than non-diabetic (35–55 years), but the difference is lesser compared to females.

The conclusions can be, first, males can have diabetes at lower age also and second, females tend to have diabetes at higher age may be due to post-menopausal hormonal disturbances. Similarly, from Fig. 2b it has been observed that average BMI of females is 28 for DM and 23 for NDM whereas for males it is 27(DM) and 25(NDM). The same is true for SBP (Fig. 2e), which infers that obese and hypertensive females have higher chance of having diabetes. There is no observable difference in HR both for males and females and also between DM and NDM subjects. Similarly, the distribution curve is very steep for RMSSD (Fig. 2g) for diabetic (males) group (between 25 and 35 ms), whereas it is distributed (between (15 and 45 ms) for non-diabetic group. But for females, RMSSD is equally distributed both for DM and NDM with higher value for diabetic. The situation is reverse for NN50

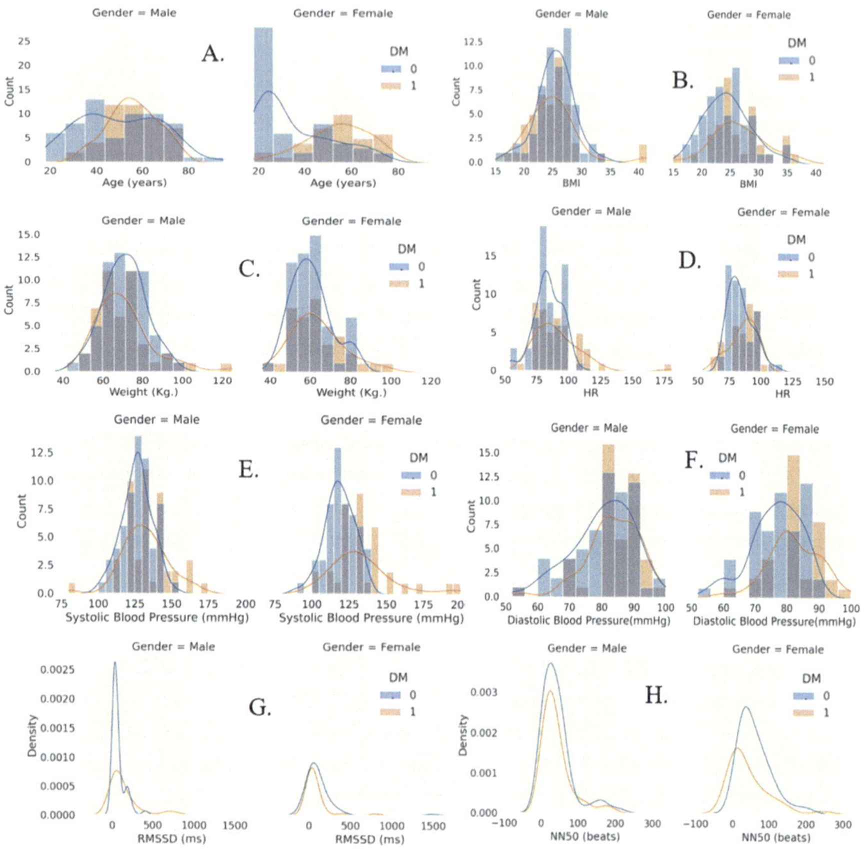

Fig. 2 Bar and density distribution plot for demographic variables and DM both for males and females for comparison study. A density plot shows the proportion of values in each range of variables. (**a**) Age vs. DM (**b**) BMI vs. DM (**c**) Weight vs. DM (**d**) HR vs. DM (**e**) SBP vs. DM (**f**) DBP vs. DM (**g**) RMSSD vs. DM (**h**) NN50 vs. DM

(Fig. 2h) where diabetic females have higher values (40–50 beats) than NDM. There is visible difference in the value of RR triangular index for females DM vs. NDM as compared to males.

3.3 *Significance Test Between all Combination of Dependent Categorical Variables and Independent Variables*

To test whether the independent variables show any significant difference for diabetic and non-diabetic group and also between men and women, the significant test was done by Student's two-tailed t-test for all possible combination of categorical dependent variables. The significance test was done at 5%, i.e., when calculated

p-value <0.05 for two-tailed t-test, the mean difference values between two categories can be considered as significant for population, not by chance. The detailed comparison is summarized in Table 4.

4 Discussion

HRV is a well-established measure to evaluate various neural and cardiac autonomic activities in the last 40 years [39–42] due to the relationship between cardiac cycle and autonomic nervous activities (sympathetic and parasympathetic). The studies for HRV are done by recording ECG signals from 24 hours to few minutes. But due to difficulty in getting good ECG recordings and its requirement for special patient preparation PRV, another technique has been used for the assessment of autonomic activities in recent years [43–47]. Also, experiencing the need for gender-based study for ANS activity assessment, many studies are done using HRV and other techniques [48–50]. Enormous studies were also done to evaluate the sex-based difference for DKD [51–53]. Inspired by these studies, this study is an effort to study sex-based differences for diabetic and control groups using PRV measures with demographic variables. PRV measures are calculated by recording FPPG signals at sitting position both for T2DM and control group subjects, and demographic and clinical data are taken through standard questionnaire.

In this study, the significant test was done for gender-based clinical and demographic characteristics of overall studied participants with conclusion that there is significant difference with p-value 0.002122 ($p < 0.05$) for diastolic blood pressure between male and female for overall population, which maybe because the average age of participated females are less than males (41 vs. 50 years). The gender-based relationship in association with DM is well summarized in Table 3 for all measured variables. It is obvious to observe that both males and females are positively correlated for age vs. DM, but females showed higher correlation ($r = 0.727$) than males ($r = 0.438$). The conclusion can be that the risk to have diabetes with age in females is higher than males. Almost the same conclusion [54, 55] is stated that T2DM is more prominent in woman with increasing age by other studies. It has been also interpreted from correlation Table 3 that BMI can be better predictor for T2DM in women than men. The similar result stating there is strong obesity-diabetes risk is associated in women [56–58]. In other study [49], the conclusion was that the women with T2DM have a higher risk of CVD associated with hypertension than men. The same is concluded from Table 3 that like aging factor females tend to have higher correlation for SBP ($r = 0.567$) and DBP ($r = 0.486$) with respect to DM than males ($r = 0.226$,SBP and $r = 0.112$,DBP).

The detailed comparison is summarized in Table 4 for significance test between all combination groups for diabetes mellitus and gender for calculated variables. The p-value (2.28668E−05) for age for DM group between men and women shows that it can be good predictor for classification job DM vs. NDM with p-value (0.3) for control groups. The same is also concluded from correlation study stated above and other studies [59, 60]. The mean value for HR (female) is 91 bpm and 78 bpm for male with p-value (1.18747E−07) for diabetic group vs. p-value (0.11) for NDM with interpretation that diabetic females have higher HR than males and hence greater risk for CVD which also stated in previous study [49]. Both males and females have higher HR with DM which is shown in Table 4 having significantly low p-value both for men and women. Similarly, both males and females with DM have higher systolic blood pressure but higher mean value for males with very low p-value (0.0003), whereas diabetic females have significant high SBP value compared to NDM females (p-value = 3.77071E−08) as compared to males (p-value = 0.03). But both values are significant as p-value <0.05, which also confirms our previous conclusion that diabetic females tend to have higher systolic blood pressure.

For PRV analysis, it is observed that except few time domain parameters, most of the measures show no significant difference between male and female. The same has been concluded by the studies on T2DM based on HRV [61, 62]. This is also true for DM and NDM subjects where some of the PRV parameters have been observed highly significant for DM and NDM subjects both for males and females and can be good predictor for diabetes. The p-value for pNN50 (%) (PRV) is 0.04 for males, and 0.01 for females that is significant ($p < 0.05$) which is also concluded by other researchers in their studies [63, 64]. Decreased RR interval and increased RR triangular index are observed [65, 66] for study between and impaired glucose metabolism for diabetic patients. But our study gave increased mean RR and RR triangular index for DM subjects with significant p-value ($0.03 < 0.05$) for DM and NDM patients. This can also be a good predictor for diabetes. The significant difference in stress index (p-value = 0.01) is observed for NDM (males and females) in comparison to DM (p-value = 0.07) subjects. It has been also found that SI for males have significant difference ($p = 0.03$) in association with DM than females ($p = 0.11$) from Table 4. This can be interpreted as men tend to develop DM because of high stress then women. Overall, it can be reported from Table 4 that there is reduction in PRV with DM both for males and females, which can also be verified by other studies done to study effect of DM on HRV [27, 67].

5 Conclusion

While HRV is a well-established measure to evaluate various neural and cardiac autonomic activities for many years, PRV has also been used widely used as a validated alternative to calculate HRV measures as time domain, frequency domain, and non-linear parameters. PRV has multiple applications for assessment and prediction objectives related to ANS activities. In this study, we tried to evaluate gender-based differences in demographic and time domain PRV parameters in association with type 2 diabetic and control groups. To measure PRV parameters, we recorded FPPG signals, which have got important benefits in terms of safety, easy to record, portability, and stable PPG waveforms compared to ECG signals. This study established that some of time domain PRV parameters can be used as DM predictors, which have significant difference in their mean values for diabetic and non-diabetic group with other demographic parameters such as age, SBP, HR, and BMI. This study also verifies that sexual dimorphism plays a key role in the pathogenesis (DM) for some demographic factors and also for PRV measures. The values generated in this study may be useful to understand and consider the different health needs and pathogenesis between women and men.

Acknowledgments We heartedly express our gratitude toward Dr. Sanjay Patil, MD Physician, Heart Sanjivani Multimedical Research Centre, Nashik, India, and Dr. Vinay Dhandania, MBBS, D. Diabetology, Diabetic Care Centre, Ranchi, India, for giving permission to record FPPG data for diabetic patients.

Conflict of Interest The authors declare no conflict of interest within themselves.

References

1. Mathur, P., Leburu, S. and Kulothungan, V., 2022. Prevalence, awareness, treatment and control of diabetes in India from the countrywide national NCD monitoring survey. *Frontiers in public health*, *10*, p.748157.
2. International Diabetes Federation. IDF Diabetes Atlas, 10th Edition. 2021. Available online:https://diabetesatlas.org/idfawp/resource_files/2021/07/IDF_Atlas_10th_Edition_2021.pdf ,accessed on 20/05/23.
3. Singer, M.E., Dorrance, K.A., Oxenreiter, M.M., Yan, K.R. and Close, K.L., 2022. The type 2 diabetes 'modern preventable pandemic'and replicable lessons from the COVID-19 crisis. *Preventive Medicine Reports*, *25*, p.101636.
4. International Institute for Population Sciences (IIPS) and ICF. National Family Health Survey (NFHS-5), 2019–2021: India. Mumbai: IIPS. 2021. Available online: http://rchiips.org/nfhs/NFHS-5_FCTS/India.pdf (accessed on 20 May 2022).
5. Anjana, R.M., Deepa, M., Pradeepa, R., Mahanta, J., Narain, K., Das, H.K., Adhikari, P., Rao, P.V., Saboo, B., Kumar, A. and Bhansali, A., 2017. Prevalence of diabetes and prediabetes in 15 states of India: results from the ICMR–INDIAB population-based cross-sectional study. *The lancet Diabetes & endocrinology*, *5*(8), pp.585-596.
6. Haffner, S.M., 2006. Relationship of metabolic risk factors and development of cardiovascular disease and diabetes. *Obesity*, *14*(S6), pp.121S-127S.
7. Smith Jr, S.C., 2007. Multiple risk factors for cardiovascular disease and diabetes mellitus. *The American journal of medicine*, *120*(3), pp.S3-S11.
8. Kilkenny, M.F., Dunstan, L., Busingye, D., Purvis, T., Reyneke, M., Orgill, M. and Cadilhac, D.A., 2017. Knowledge of risk factors for diabetes or cardiovascular disease (CVD) is poor among individuals with risk factors for CVD. *PloS one*, *12*(2), p.e0172941.
9. Singh, G.M., Danaei, G., Farzadfar, F., Stevens, G.A., Woodward, M., Wormser, D., Kaptoge, S., Whitlock, G., Qiao, Q., Lewington, S. and Di Angelantonio, E., 2013. The age-specific quantitative effects of metabolic risk factors on cardiovascular diseases and diabetes: a pooled analysis. *PloS one*, *8*(7), p.e65174.
10. Torgerson, J.S., Hauptman, J., Boldrin, M.N. and Sjöström, L., 2004. XENical in the Prevention of Diabetes in Obese Subjects (XENDOS) Study. A randomized study of orlistat as an adjunct to lifestyle changes for the prevention of type 2 diabetes in obese patients. *Clinical Diabetology*, *5*(2), pp.95-104.

11. Karlsson, J., Taft, C., Sjöström, L., Torgerson, J.S. and Sullivan, M., 2003. Psychosocial functioning in the obese before and after weight reduction: construct validity and responsiveness of the Obesity-related Problems scale. *International journal of obesity*, *27*(5), pp.617-630.
12. Muilwijk, M., Bolijn, R., Galenkamp, H., Stronks, K., van Charante, E.M. and van Valkengoed, I.G., 2022. The association between gender-related characteristics and type 2 diabetes risk in a multi-ethnic population: The HELIUS study. *Nutrition, Metabolism and Cardiovascular Diseases*, *32*(1), pp.142-150.
13. Kautzky-Willer, A., Harreiter, J. and Pacini, G., 2016. Sex and gender differences in risk, pathophysiology and complications of type 2 diabetes mellitus. *Endocrine reviews*, *37*(3), pp.278-316.
14. Anand, S.S., Islam, S., Rosengren, A., Franzosi, M.G., Steyn, K., Yusufali, A.H., Keltai, M., Diaz, R., Rangarajan, S. and Yusuf, S., 2008. Risk factors for myocardial infarction in women and men: insights from the INTERHEART study. *European heart journal*, *29*(7), pp.932-940.
15. Peters, S.A., Huxley, R.R. and Woodward, M., 2014. Diabetes as risk factor for incident coronary heart disease in women compared with men: a systematic review and meta-analysis of 64 cohorts including 858,507 individuals and 28,203 coronary events. *Diabetologia*, *57*, pp.1542-1551.
16. De Hauteclocque, A., Ragot, S., Slaoui, Y., Gand, E., Miot, A., Sosner, P., Halimi, J.M., Zaoui, P., Rigalleau, V., Roussel, R. and Saulnier, P.J., 2014. The influence of sex on renal function decline in people with Type 2 diabetes. *Diabetic Medicine*, *31*(9), pp.1121-1128.
17. Gómez-Marcos, M.Á., Recio-Rodríguez, J.I., Gomez-Sanchez, L., Agudo-Conde, C., Rodriguez-Sanchez, E., Maderuelo-Fernandez, J., Gomez-Sanchez, M. and Garcia-Ortiz, L., 2015. Gender differences in the progression of target organ damage in patients with increased insulin resistance: the LOD-DIABETES study. *Cardiovascular diabetology*, *14*(1), pp.1-11.
18. Lin, C.H., Yang, W.C., Tsai, S.T., Tung, T.H. and Chou, P., 2007. A community-based study of chronic kidney disease among type 2 diabetics in Kinmen, Taiwan. *Diabetes research and clinical practice*, *75*(3), pp.306-312.
19. McCollum, M., Hansen, L.B., Lu, L. and Sullivan, P.W., 2005. Gender differences in diabetes mellitus and effectson self-care activity. *Gender medicine*, *2*(4), pp.246-254.
20. Klimek, P., Kautzky-Willer, A., Chmiel, A., Schiller-Frühwirth, I. and Thurner, S., 2015. Quantification of diabetes comorbidity risks across life using nation-wide big claims data. *PLoS computational biology*, *11*(4), p.e1004125.
21. Huikuri, H.V., Pikkuja¨ msa¨, S.M., Airaksinen, K.J., Ika¨ heimo, M.J., Rantala, A.O., Kauma, H., Lilja, M. and Kesa¨ niemi, Y.A., 1996. Sex-related differences in autonomic modulation of heart rate in middle-aged subjects. *Circulation*, *94*(2), pp.122-125.
22. Jensen-Urstad, K., Storck, N., Bouvier, F., Ericson, M., Lindbland, L.E. and Jensen-Urstad, M., 1997. Heart rate variability in healthy subjects is related to age and gender. *Acta Physiologica Scandinavica*, *160*(3), pp.235-241.
23. Kuo, T.B., Lin, T., Yang, C.C., Li, C.L., Chen, C.F. and Chou, P., 1999. Effect of aging on gender differences in neural control of heart rate. *American Journal of Physiology-Heart and Circulatory Physiology*, *277*(6), pp.H2233-H2239.
24. Umetani, K., Singer, D.H., McCraty, R. and Atkinson, M., 1998. Twenty-four hour time domain heart rate variability and heart rate: relations to age and gender over nine decades. *Journal of the American College of Cardiology*, *31*(3), pp.593-601.
25. Liao, D., Cai, J., Brancati, F.L., Folsom, A., Barnes, R.W., Tyroler, H.A. and Heiss, G., 1995. Association of vagal tone with serum insulin, glucose, and diabetes mellitus—The ARIC Study. *Diabetes research and clinical practice*, *30*(3), pp.211-221.
26. Nolan, R.P., Barry-Bianchi, S.M., Mechetiuc, A.E. and Chen, M.H., 2009. Sex-based differences in the association between duration of type 2 diabetes and heart rate variability. *Diabetes and Vascular Disease Research*, *6*(4), pp.276-282.
27. Benichou, T., Pereira, B., Mermillod, M., Tauveron, I., Pfabigan, D., Maqdasy, S. and Dutheil, F., 2018. Heart rate variability in type 2 diabetes mellitus: A systematic review and meta–analysis. *PloS one*, *13*(4), p.e0195166.

28. Lan, K.C., Raknim, P., Kao, W.F. and Huang, J.H., 2018. Toward hypertension prediction based on PPG-derived HRV signals: A feasibility study. *Journal of medical systems*, *42*, pp.1-7.
29. Castaldo, R., Melillo, P., Bracale, U., Caserta, M., Triassi, M. and Pecchia, L., 2015. Acute mental stress assessment via short term HRV analysis in healthy adults: A systematic review with meta-analysis. *Biomedical Signal Processing and Control*, *18*, pp.370-377.
30. Piotrowski, Z. and Szypulska, M., 2017. Classification of falling asleep states using HRV analysis. *Biocybernetics and biomedical engineering*, *37*(2), pp.290-301.
31. Castaldo, R., Montesinos, L., Melillo, P., Massaro, S. and Pecchia, L., 2018. To What Extent Can We Shorten HRV Analysis in Wearable Sensing? A Case Study on Mental Stress Detection. In *EMBEC & NBC 2017: Joint Conference of the European Medical and Biological Engineering Conference (EMBEC) and the Nordic-Baltic Conference on Biomedical Engineering and Medical Physics (NBC), Tampere, Finland, June 2017* (pp. 643-646). Springer Singapore.
32. Mejía-Mejía, E., May, J.M., Torres, R. and Kyriacou, P.A., 2020. Pulse rate variability in cardiovascular health: A review on its applications and relationship with heart rate variability. *Physiological Measurement*, *41*(7), p.07TR01.
33. Allen, J., 2007. Photoplethysmography and its application in clinical physiological measurement. *Physiological measurement*, *28*(3), p.R1.
34. https://www.who.int/publications-detail-redirect/use-of-glycated-haemoglobin-(-hba1c)-in-diagnosis-of-diabetes-mellitus (As on 28-09-2023)
35. Yu, S.G., Kim, S.E., Kim, N.H., Suh, K.H. and Lee, E.C., 2021. Pulse rate variability analysis using remote photoplethysmography signals. *Sensors*, *21*(18), p.6241.
36. Electrophysiology, T.F.O.T.E.S.O.C.T.N.A.S.O.P., 1996. Heart rate variability: standards of measurement, physiological interpretation, and clinical use. *Circulation*, *93*(5), pp.1043-1065.
37. Shaffer, F. and Ginsberg, J.P., 2017. An overview of heart rate variability metrics and norms. *Frontiers in public health*, p.258.
38. Baevsky, R.M. and Berseneva, A.P., 2009. Use KARDiVAR system for determination of the stress level and estimation of the body adaptability: Standards of measurements and physiological interpretation. *Kardivar. TV.* https://pdfs.semanticscholar.org/74a2/92bf afca4fdf, *1149*, p.d557.
39. Xhyheri, B., Manfrini, O., Mazzolini, M., Pizzi, C. and Bugiardini, R., 2012. Heart rate variability today. *Progress in cardiovascular diseases*, *55*(3), pp.321-331.
40. Quintana, D.S., 2017. Statistical considerations for reporting and planning heart rate variability case-control studies. *Psychophysiology*, *54*(3), pp.344-349.
41. Huikuri, H.V., Mäkikallio, T., Airaksinen, K.J., Mitrani, R., Castellanos, A. and Myerburg, R.J., 1999. Measurement of heart rate variability: a clinical tool or a research toy?. *Journal of the American College of Cardiology*, *34*(7), pp.1878-1883.
42. Malik, M., Huikuri, H., Lombardi, F., Schmidt, G. and e-Health/Digital Rhythm Study Group of the European Heart Rhythm Association, 2017. The purpose of heart rate variability measurements. *Clinical Autonomic Research*, *27*, pp.139-140.
43. Ahmad, S., Bolic, M., Dajani, H. and Groza, V., 2009, May. Wavelet estimation of pulse rate variability from oscillometric blood pressure measurements. In *2009 IEEE International Workshop on Medical Measurements and Applications* (pp. 37-40). IEEE.
44. Constant, I., Laude, D., Murat, I. and Elghozi, J.L., 1999. Pulse rate variability is not a surrogate for heart rate variability. *Clinical Science*, *97*(4), pp.391-397.
45. Gil, E., Orini, M., Bailon, R., Vergara, J.M., Mainardi, L. and Laguna, P., 2010. Photoplethysmography pulse rate variability as a surrogate measurement of heart rate variability during non-stationary conditions. *Physiological measurement*, *31*(9), p.1271.
46. Jeyhani, V., Mahdiani, S., Peltokangas, M. and Vehkaoja, A., 2015, August. Comparison of HRV parameters derived from photoplethysmography and electrocardiography signals. In *2015 37th annual international conference of the ieee engineering in medicine and biology society (EMBC)* (pp. 5952-5955). IEEE.
47. Rauh, R., Limley, R., Bauer, R.D., Radespiel-Troger, M. and Mueck-Weymann, M., 2004, August. Comparison of heart rate variability and pulse rate variability detected with pho-

toplethysmography. In *Saratov Fall Meeting 2003: Optical Technologies in Biophysics and Medicine V* (Vol. 5474, pp. 115-126). SPIE.
48. Huikuri, H.V., Pikkuja¨ msa¨, S.M., Airaksinen, K.J., Ika¨ heimo, M.J., Rantala, A.O., Kauma, H., Lilja, M. and Kesa¨ niemi, Y.A., 1996. Sex-related differences in autonomic modulation of heart rate in middle-aged subjects. *Circulation*, *94*(2), pp.122-125.
49. Succurro, E., Miceli, S., Fiorentino, T.V., Sciacqua, A., Perticone, M., Andreozzi, F. and Sesti, G., 2021. Sex-specific differences in left ventricular mass and myocardial energetic efficiency in non-diabetic, pre-diabetic and newly diagnosed type 2 diabetic subjects. *Cardiovascular Diabetology*, *20*(1), pp.1-11.
50. Fagot-Campagna, A., Saaddine, J., Narayan, K.V., Goldschmid, M. and Howard, B.V., 1999. Re:"Sex differences in risk factors for clinical diabetes mellitus in a general population: a 12-year follow-up of the Finnmark study". *American journal of epidemiology*, *149*(11), pp.1073-1074.
51. Maric-Bilkan, C., 2020, March. Sex differences in diabetic kidney disease. In *Mayo Clinic Proceedings* (Vol. 95, No. 3, pp. 587-599). Elsevier.
52. García, G.G., Iyengar, A., Kaze, F., Kierans, C., Padilla-Altamira, C. and Luyckx, V.A., 2022, March. Sex and gender differences in chronic kidney disease and access to care around the globe. In *Seminars in nephrology* (Vol. 42, No. 2, pp. 101-113). WB Saunders.
53. Piani, F., Melena, I., Tommerdahl, K.L., Nokoff, N., Nelson, R.G., Pavkov, M.E., van Raalte, D.H., Cherney, D.Z., Johnson, R.J., Nadeau, K.J. and Bjornstad, P., 2021. Sex-related differences in diabetic kidney disease: A review on the mechanisms and potential therapeutic implications. *Journal of diabetes and its complications*, *35*(4), p.107841.
54. Relimpio, F., Martinez-Brocca, M.A., Leal-Cerro, A., Losada, F., Mangas, M.A., Pumar, A. and Astorga, R., 2004. Variability in the presence of the metabolic syndrome in type 2 diabetic patients attending a diabetes clinic: influences of age and gender. *Diabetes research and clinical practice*, *65*(2), pp.135-142.
55. Kuk, J.L. and Ardern, C.I., 2010. Age and sex differences in the clustering of metabolic syndrome factors: association with mortality risk. *Diabetes care*, *33*(11), pp.2457-2461.
56. Kuk, J.L., Lee, S., Heymsfield, S.B. and Ross, R., 2005. Waist circumference and abdominal adipose tissue distribution: influence of age and sex. *The American journal of clinical nutrition*, *81*(6), pp.1330-1334.
57. Wannamethee, S.G., Papacosta, O., Whincup, P.H., Carson, C., Thomas, M.C., Lawlor, D.A., Ebrahim, S. and Sattar, N., 2010. Assessing prediction of diabetes in older adults using different adiposity measures: a 7 year prospective study in 6,923 older men and women. *Diabetologia*, *53*, pp.890-898.
58. Meisinger, C., Döring, A., Thorand, B., Heier, M. and Löwel, H., 2006. Body fat distribution and risk of type 2 diabetes in the general population: are there differences between men and women? The MONICA/KORA Augsburg cohort study. *The American journal of clinical nutrition*, *84*(3), pp.483-489.
59. Becker, J., Nora, D.B., Gomes, I., Stringari, F.F., Seitensus, R., Panosso, J.S. and Ehlers, J.A.C., 2002. An evaluation of gender, obesity, age and diabetes mellitus as risk factors for carpal tunnel syndrome. *Clinical Neurophysiology*, *113*(9), pp.1429-1434.
60. Bierhaus, A., Hofmann, M.A., Ziegler, R. and Nawroth, P.P., 1998. AGEs and their interaction with AGE-receptors in vascular disease and diabetes mellitus. I. The AGE concept. *Cardiovascular research*, *37*(3), pp.586-600.
61. Giordano, M., Manzella, D., Paolisso, G., Caliendo, A., Varricchio, M. and Giordano, C., 2001. Differences in heart rate variability parameters during the post-dialytic period in type II diabetic and non-diabetic ESRD patients. *Nephrology Dialysis Transplantation*, *16*(3), pp.566-573.
62. Rasic-Milutinovic, Z.R., Milicevic, D.R., Milovanovic, B.D., Perunicic-Pekovic, G.B. and Pencic, B.D., 2010. Do components of metabolic syndrome contribute to cardiac autonomic neuropathy in non-diabetic patients. *Saudi Med J*, *31*(6), pp.650-657.

63. Molon, G., Costa, A., Bertolini, L., Zenari, L., Arcaro, G., Barbieri, E. and Targher, G., 2007. Relationship between abnormal microvolt T-wave alternans and poor glycemic control in type 2 diabetic patients. *Pacing and Clinical Electrophysiology*, *30*(10), pp.1267-1272.
64. Boer-Martins, L., Figueiredo, V.N., Demacq, C., Martins, L.C., Consolin-Colombo, F., Figueiredo, M.J., Cannavan, F.P. and Moreno, H., 2011. Relationship of autonomic imbalance and circadian disruption with obesity and type 2 diabetes in resistant hypertensive patients. *Cardiovascular diabetology*, *10*(1), pp.1-12.
65. Giordano, M., Manzella, D., Paolisso, G., Caliendo, A., Varricchio, M. and Giordano, C., 2001. Differences in heart rate variability parameters during the post-dialytic period in type II diabetic and non-diabetic ESRD patients. *Nephrology Dialysis Transplantation*, *16*(3), pp.566-573.
66. Wang G. Effects of impaired glucose metabolism on heart rate variability and blood pessure variability in essential hpertensive patients. J Huazhong Univ Sci Technol. 2006; 26: 654±656.
67. Stuckey, M. and Petrella, R.J., 2013. Heart rate variability in type 2 diabetes mellitus. Critical Reviews™ in Biomedical Engineering, 41(2).

Symbiotic Organism Search: A Novel Paradigm for Feature Optimization in Finger Vein Recognition Systems

P. Jayapriya and K. Umamaheswari

1 Introduction

Biometric are significantly used in person authentication. Biometrics is more dependable and simple to use than traditional methodologies. The identification of individuals based on their physical traits is known as biometric authentication. The hand-based biometrics such as finger print, finger knuckle, palmprint, and hand geometry are external to human and are vulnerable to spoofing and forgery. Researchers are especially interested in the finger vein biometric because of its low device restriction, forgery, threat, liveliness detection, and high accuracy. Finger vein recognition is a technique for identifying a person based on the vein pattern in their fingers. Vein patterns appear as a series of black lines due to deoxyhemoglobin in the blood absorbs near-infrared light [1]. Finger vein biometrics is more dependable and trustworthy than traditional biometrics. The feature extraction strategy used in most of the current available approaches for finger vein recognition [2–4] is similar in that it uses features from the segmented blood vascular network for recognition. The finger vein pictures, however, are not always clear and sometimes reveal uneven shadings due to optical blurring and skin scattering issues [5]. As a result of the low quality of finger vein images, segmentation mistakes can occur throughout the feature extraction procedure. The recognition accuracy may be harmed if the networks are not segregated adequately. To solve this problem, [6] proposed a method for finger vein recognition Histogram of Oriented Gradients (HOG) and Gray Level Co-occurrence Matrix (GLCM). The captured finger vein

P. Jayapriya (✉)
Sri Eshwar College of Engineering, Coimbatore, Tamil Nadu, India

K. Umamaheswari
PSG College of Technology, Coimbatore, Tamil Nadu, India

A. Haldorai et al. (eds.), *6th EAI International Conference on Big Data Innovation for Sustainable Cognitive Computing*, EAI/Springer Innovations in Communication and Computing, https://doi.org/10.1007/978-3-031-54696-9_8

images are enhanced by CALHE in the proposed method, and then HOG and GLCM are used to extract features from the enhanced images. Nearest Neighbor is used to calculate the similarity between extracted and enrolled features.

One of the most difficult challenges in a hand biometric system is feature selection. It entails determining which variables are the most important and whether they provide enough information to perform an efficient classification operation. Although a huge number of hand features can be extracted, this does not guarantee the best accuracy because many of them may be unnecessary or even meaningless, and the classifier may not be able to generalize effectively.

Almost all metaheuristic algorithms have the same characteristics: they are inspired by nature, they employ random variables, they don't require much gradient information, and they contain multiple parameters that must be suited to the situation at hand [7]. Each metaheuristic method has distinct advantages in terms of robustness and performance in noisy settings, with uncertain parameters, and across diverse problem spaces [8]. However, according to the "no-free-lunch" theorem, no single metaheuristic algorithm can solve all optimization problems optimally [9].

The SOS algorithm, developed by Cheng and Prayogo in 2014, is one of the most successful and enticing meta-heuristic algorithms inspired by the communication of organisms in nature [10]. In this work, Symbiotic Organisms Search is a new simple and powerful metaheuristic algorithm introduced. This optimization algorithm mimics organisms' symbiotic interaction techniques for surviving in the ecosystem. This method computes three distinct and chain processes known as commensalism, mutualism, and parasitic life to improve population solutions while presenting no parameters to restrict the two actions of exploration and exploitation [11]. Most optimization problems have been successfully solved as a result of this. The SOS algorithm has a significant benefit over most other metaheuristic algorithms in that it requires no specific algorithm parameters to perform operations.

The remaining part of the chapter is structured as follows: Sect. 2 is a literature review, Sect. 3 is a proposed SOS feature selection methodology for finger vein biometrics, Sect. 4 is a discussion of the results, and Sect. 5 is the conclusion.

2 Literature Study

The several strategies have been implemented by experts in the field of biometric authentication. This section of the literary works presents an overview of some significant contributions to the existing literature.

Xi et al. [12] proposed Pyramid Histograms of Gray, Texture and Orientation Gradients (PHGTOG) which extracts the global spatial information and local texture and shape features. In addition to improve the performance and to reduce the computational complexity, the optimal subset of features are selected using sparse weight vector which is trained by LASSO. The experimental results show better performance for the selected features.

Wu et al. [13] proposed an image preprocessing with classification for finger vein recognition system. The preprocessing is done by Principle Component Analysis (PCA) and Linear Discriminant Analysis (LDA) feature extraction techniques. Both the LDA and PCA are dimension reduction algorithms which reduces the feature vector. The classification of these features is done by Support Vector Machine (SVM) with adaptive neuro fuzzy inference system (ANFIS). The experimental analysis shows that the SVM classification achieves 98% accuracy which shows better than the existing system artificial neural network of ANFIS.

Rosdi et al. [14] developed the local line binary pattern (LLBP) method for extracting features for finger vein images. The LLPB method is an improved variant of the LBP (Local binary pattern) approach that extracts features in both horizontal and vertical directions. The line patterns obtained just from vertical and horizontal finger vein images have significant orientation information, while the line patterns acquired exclusively from vertical and horizontal finger vein images have rich orientation information.

Zhang and Wang [15] proposed a GLCM_HOG feature extraction technique. The image enhancement is done by CLHE and adaptive median filtering used to improve the texture of the finger vein image. Next features are extracted using HOG_GLCM feature extraction approach. The Principle Component Analysis (PCA) is applied to the resultant feature vector to reduce the dimensionality of the feature vector. Finally, the classification accuracy is based on SVM achieves 97.3%. Here considerably the speed of the recognition is improved.

Yang et al. [16] developed an integration of indexing method for finger vein pattern to improve the accuracy and efficiency. The indexing is done with the extracted veins patterns directions and is represented by elliptical map. The similarity of the vein directions are measured by hamming distance. Here the same direction of veins is considered for matching. The experimental results show that the integration work gives improved results.

Kauba et al. [17] suggested the feature level fusion approach to improve the unimodal finger vein recognition accuracy. The feature extraction techniques such as maximum curvature, repeated line tracking wide line detector are used to extract the finger vein features. The various fusion approaches such as majority voting, weighted average STAPLE are implemented. The experimental result shows that multi-algorithm feature extraction at feature level fusion improves the recognition accuracy of the unimodal finger vein biometric recognition.

Yang et al. [18] also suggested that multi-feature extraction improves the performance of the unimodal biometric recognition. The author proposed a new feature-level fusion of global and local features. The Gabor filter is used to extract the local feature and moment invariant for global features. And these features are fused based on weighted fusion approach. The classification accuracy based on nearest neighbor shows better results for finger vein personal authentication.

The fusion of different features improves the performance of the biometric recognition. Lu et al. [19] propose the finger vein recognition based on score level fusion of Gabor features. To improve the feature of the image, the preprocessing is done using contrast limited adaptive histogram equalization. The eight channels of the Gabor features are extracted and weight matching score level fusion rule is applied to find the matching score of the image. The experimental results show better performance in finger vein recognition.

According to the literature, extracting texture features using a multi-feature extraction technique results in a higher recognition rate. The problem is that using multiple feature extraction techniques results in a larger number of feature vectors. The larger size vector thus slows down the classification process and reduces the accuracy. To address these difficulties, multi-feature extraction techniques such as HOG and GLCM are used to extract the textural features of the finger vein. The optimal subset of features is then determined using the proposed bio-inspired optimization algorithm termed Symbiotic Organization Search (SOS) to reduce computing cost and maximize accuracy.

3 A Proposed SOS Feature Selection Methodology for Finger Vein Recognition

To improve finger vein recognition classification and identification accuracy, the proposed approach uses a metaheuristic algorithm for Feature Selection (FS) to obtain a reduced number of feature subsets. The present study includes preprocessing, feature extraction, feature subset selection using Symbiotic Organism Search (SOS), and classification. Figure 1 depicts the block diagram of the proposed methodology.

3.1 Preprocessing

The sensor-captured image of a finger vein is pre-processed and used as input for the next feature extraction phase [20, 21]. The image may be affected during the finger vein acquisition due to variations in rotation and translation, as well as lighting. The performance of the finger vein recognition will suffer as a result. To address these issues, the Region of Interest (ROI) is extracted using canny edge detection and then the image background is removed. Following ROI extraction, the image is enhanced with histogram equalization [22] to increase the intensity of grey level contrast in the finger vein image.

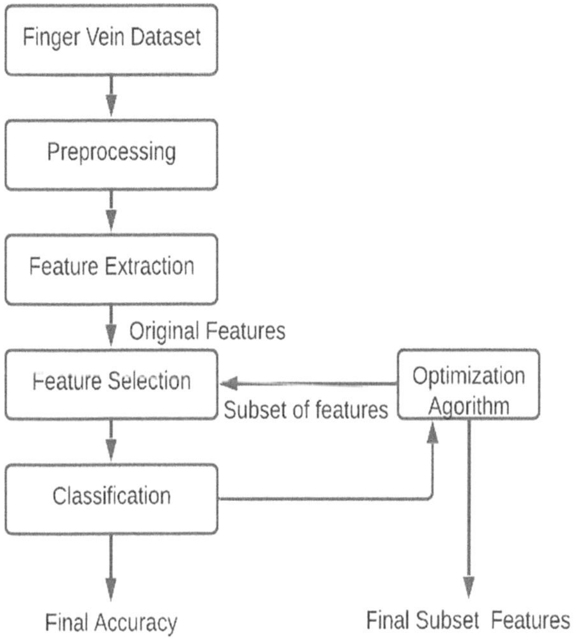

Fig. 1 Architecture of the proposed SOS feature: Selection for FV recognition system

3.2 *Feature Extraction*

Following the preprocessing of the FV images, the features are extracted from the image using feature extraction techniques. The texture-based feature extraction techniques HOG and GLCM are used in this work to extract features from FV images. The detailed examination of the feature descriptors is provided below.

3.2.1 Histogram of Oriented Gradients

The texture and shape of an image are extracted using the HOG feature extraction approach. HOG is also used to extract the edge direction based on the image's magnitude and orientation. This technique captures the image's localized portions. The image is divided into smaller regions, and HOG generates histograms for each of these regions separately. The histogram is generated using the image pixel's gradients and orientations [23].

The advantage of using the HOG descriptor is it overcomes the illumination, rotational, and translational variation of the image. First the gradient of the image is calculated for the image (I) as (x, y) and is mathematically expressed as follows:

$$G_x = N \times I(x,y) \text{ and } G_y = N^T \times I(x,y) \tag{1}$$

G_x and G_y represent the vertical and horizontal directions of the gradient, N represents gradient operator, and N^T indicates the gradient transformation. The gradient and edge of the x and y are calculated using Eqs. (2) and (3)

$$G(x,y) = \sqrt{\left(G_x(x,y)\right)^2 + \left(G_y(x,y)\right)^2} \tag{2}$$

$$\theta(x,y) = \tan^{-1} \frac{G_y(x,y)}{G_x(x,y)} \tag{3}$$

The calculated histogram values are normalized to avoid the noise and to improve the illumination of the finger vein image.

3.2.2 Gray Level Occurrence Matrix

GLCM is a second order statics used to extract the texture features of the given image. GLCM calculates the distance between x and y gray level pair. The gray level intensity pixel value of x is calculated for particular pixel in y. The GLCM size depends on the gray level pair(x, y). It also reveals the certain features of gray level spatial space. The shape of the object decides the features of the image texture. We use the textural feature extraction based on Haralick [24, 25]. The co-occurrence of each image is calculated by $G(x, y)$ with distance d and Orientation θ. For the given set of features, thirteen Haralick features are constructed for the given finger vein image.

3.3 *Feature Selection*

The finger vein is rich in texture features, and the hybrid HOG2LCM is used to extract the most distinguishing texture features. The hybrid HOG2LCM gives a larger number of features for pattern classification than samples, resulting in over fitting. Insignificant features such as irrelevancy, duplication, and noise may be included in the data set if the number of input variables is too great, but dependent features may be omitted. Therefore, the relevant features are selected based on the proposed new metaheuristic optimization technique known as Symbiotic Organization Search (SOS). The reason behind the selection of this optimization technique is it does not contain the specific parameters like other metaheuristic algorithms. Therefore, the SOS algorithm will not suffer from high computational time and convergence to local optima solution due to inappropriate tuning of parameters.

However, unlike most other metaheuristic algorithms, SOS algorithm has not any specific algorithm parameters [26]. Therefore, problems of increasing computational time and convergence to the local optimal solution arising from unsuitable tuning of the parameters are not seen in SOS. It only includes general parameters such as population size and maximum iteration number. The algorithm consists of mutualism phase, commensalism phase, and parasitism phase which are biological interactions seen most in the real world.

The optimal feature subset for finger vein biometric is selected using a novel approach based on SOS. In SOS, each feature subset can be seen as a position of an organism. The original subset contains N number of features. The number of features is reduced to optimal subset of features, which gives better classification accuracy. Each subset of features is evaluated according to the fitness function. Here, the fitness function in terms of KNN classification accuracy is used to validate the subset of attributes.

The Symbiotic Organism Search (SOS) method is a new population-based metaheuristic [27]. SOS comprises of three stages that reflect the relationship of different species in the eco system: mutualism, commensalism, and parasitism. Mutualism occurs when both species benefit from the association. Commensalism occurs when one person benefits while another suffers no harm. Parasitism occurs when one species benefits while another suffers. Each organism member is a vector, and each organism in the search space is assigned a value. The algorithm iteratively populates all possible solutions to obtain the global optimum solution vector. The formulation of the standard SOS is outlined as shown in Algorithm 1.

3.3.1 Mutualism

In mutualism, the two organisms (M_i, M_j) benefited each other and the mathematical expression is as follows

$$M_i^{\text{new}} = M_i + rand(0,1) \times \left(X_{\text{best}} - \text{MV} \times \text{BF}_1\right) \tag{4}$$

$$M_j^{\text{new}} = M_j + rand(0,1) \times \left(X_{\text{best}} - \text{MV} \times \text{BF}_2\right) \tag{5}$$

where MV is mutual vector and BF is a beneficial factor.

The MV and BF are calculated using the following equations:

$$\text{MV} = \frac{M_i + M_j}{2} \tag{6}$$

$$\text{BF}_1 = 1 + \text{round}\left(rand(0,1)\right) \tag{7}$$

$$\text{BF}_2 = 1 + \text{round}\left(rand(0,1)\right) \tag{8}$$

The structure of mutual vector and beneficial vector is changed to get the best organism of the present population X^{best}. The mutual vector represents the relationship between two organisms and BF represents the beneficial achievements of the two organisms.

3.3.2 Commensalism

In this phase, the one organism is allowed to increase the fitness value. The organism M_i is chosen randomly for each organism in M_i. The mathematical equation to illustrate the commensalism between the two organisms is shown in Eq. (9)

$$M_i^{new} = M_i + rand(-1,1) \times (M_{best} - M_j) \tag{9}$$

The beneficial impact of organism M_j is provided to organism M_i. Here the newly generated population is replaced only with predecessor M_i with condition that if its fittness is better.

3.3.3 Parasitism

The final stage mimics the concept of parasites intending to completely eradicate their hosts. The same concept is used in the algorithm, where the parasite, known as a Parasite Vector (PV), is created by modifying organism M_i in such a way that it deviates from the original organism. This parasite remains to eliminate its host, and the role is played by a randomly selected organism M_j. One possible scenario is that the host can still survive if its fitness is superior to that of the parasite and the parasite is removed instead.

The SOS first initializes the ecosystem and is expressed as X = {X!, X@,... Xecoszie}. Furthermore, as mentioned in [28], the population size of the organisms can be chosen between a moderate population size of 25 and a large population size of 100. By comparing the respective objective function values to the best object function value, the best one is chosen as X best. The process is repeated until the global best solution is learned. When the maximum number of fitness function evaluations is reached, the process is terminated. Otherwise, the algorithm keeps evaluating by exploring and exploiting new search areas for solutions. Reducing the duration of the algorithm execution time, for example, may not yield solutions that are close to the desired global optimum, while lengthening the execution time may raise the algorithm's computational cost [27]. Figure 2 shows the proposed SOS work flow.

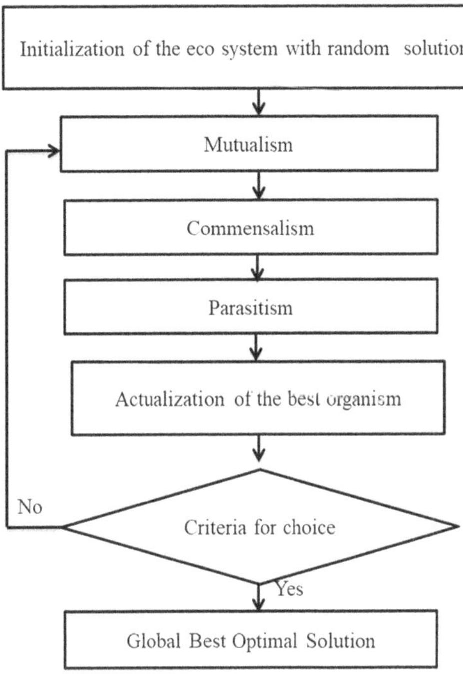

Fig. 2 Proposed SOS methodology work flow

Algorithm 1: Outline of Symbiotic Organisms Search

```
Initialize the population size N, maximum iterations 100,
Upper bound (UB), Lower Bound (LB),
Identify the best feature vector in initial population (M_best)
Loop:
While iteration < max iterations
For i=1: N
//Mutualism Phase
Select organisms (feature vector) M_i, M_j
Calculate MV using Eq. (6)
Calculate BF using Eqs. (7) and (8)
Generate new organism M_i^new, M_j^new based on Eqs. (4) and (5)
Replace predecessors when the fitness is better for new organism

// Commensalism
Select organisms (feature vector) M_i, M_j
Generate new organism M_i^new, M_j^new based on Eq. (9)
Replace predecessors when the fitness is better for new organism

//Parasitism
Select organisms (feature vector) M_i, M_j
```

```
 Modify the Mi and generate the PV
 Replace Mi when the fitness is better than PV
 End For

 Update best Organisms (Feature vector) Mbest for the present
population
 End while
```

3.3.4 KNN Classifier

For suitable values of k, the k-nearest neighbor classifier is a simple supervised classifier that produces good results. This classifier calculates the distance between each training data point and the testing dataset, then chooses the best k neighbors with the least distance. The lack of a training phase simplifies the execution of this classifier. The Euclidean distance is utilized as the distance metric in this study, and $k = 1$ is the best value found after the process of trial and error.

4 Result and Discussion

The experimental analysis of finger vein recognition in this work is based on a multi-algorithm feature extraction and metaheuristic feature selection technique. The proposed SOS feature selection algorithm and the hybrid HOG2LCM for feature extraction are implemented to choose the best subset of features. To classify the accuracy of the finger vein recognition system, the KNN is employed as a classifier. The experiment was run on a Windows 7 Intel Core i7 CPU with a 2.53 GHz operating frequency and 8GB of available memory, using MATLAB R2018. The data for this study was collected from 35 people, each of whom had five fingers veins, yielding 175 samples. Three sessions of four images each were taken, with at least 1 week between each. This section presents the results of the proposed techniques for each dataset individually.

Tables 1 and 2 depict the optimal subset of features and recognition accuracy using the SOS feature selection for THU_FVFDT and FV-USM Datasets. As the number of generations increases, the classification accuracy for both feature selection algorithms improves. In every evaluated feature, the accuracy of THU_FVFDT and FV-USM Datasets is comparable but with a reduced number of features. The THU_FVFDT and FV-USM Datasets feature selection algorithms have an accuracy

Table 1 Performance of the classification accuracy for proposed feature selection based on FV-USM database

Feature extraction	Total features	Optimized features	Accuracy
Hybrid HOG^{2}LCM	620	620	97.2
HOG^{2}LCM features with proposed SOS	620	102	98.67

Table 2 Comparison of the proposed feature selection algorithms based on number of generations

No of generations	Input HOG²LCM features (A)	Reduced No of features (A-B)		Optimized selected features (B)		Accuracy (%)	
		THU-FVFDT	FV-USM D	THU-FVFDT	FV-USM D	THU-FVFDT	FV-USM D
10	620	586	570	34	50	83.7	80
15	620	518	566	102	54	89.3	82.3
20	620	527	527	93	93	93.4	87.5
25	620	555	522	65	98	97.2	90
30	620	566	518	54	102	98.7	98.67

of 98.7% and 98.2%, respectively. Besides, for 30 runs, the THU_FVFDT database has 54 selected features, while FV-USM database has 102 features.

Table 2 shows the results for THU-FVFDT and FV-USM database based on the number of generations. The results show that THU-FVFDT database shows 98.7% classification accuracy with 54 features. Similarly the FV-USM database shows 98.67% accuracy with 102 features.

5 Conclusion

In this chapter, a finger-vein pattern recognition system based on hybrid HOG²LCM for feature extraction and Symbiotic Organism Search (SOS) optimization technique for feature selection and classification using a KNN are proposed. It is well known if the dimension of the input data is large enough, it would cause overload of the system. The proposed SOS optimization technique is used to select the most distinguishing feature vectors and then classify them with KNN. In the FV identification system, the proposed SOS is compared to classic PSO for identifying significant features. The experimental results showed the proposed system to be an effective method for FV biometric authentication. The identification rate of proposed SOS feature selection shows superior performance with reduced number of features for personal identification. The finger-vein biometric recognition system is more secure and reliable than prior methods of person identification, but it might be used in general infrastructure.

References

1. E. Ting and M.Z. Ibrahim, "A review of finger vein recognition system," Journal of Telecommunication, Electronic and Computer Engineering, 10(1–9), pp. 167–17, 2018
2. N. Miura, A. Nagasaka and T. Miyatake, "Feature extraction of finger-vein patterns based on repeated line tracking and its application to personal identification," Machine vision and applications, vol 15, pp. 194–203, Oct 2004.
3. C.B Yu, H.F Qin, L. Zhang and Y.Z. Cui, "Finger-vein image recognition combining modified hausdorff distance with minutiae feature matching," Interdisciplinary Sciences: Computational Life Sciences, vol 1, pp. 280–289, 2009.

4. Song, W, Kim, T, Kim, H.C, Choi, J.H, Kong, H.J and S. R Lee, "A finger-vein verification system using mean curvature," Patt. Recogn. Lett, vol 32, pp. 1541–1547, 2011.
5. E.C. Lee and K. R Park, "Image restoration of skin scattering and optical blurring for finger vein recognition. Opt. Lasers Eng," vol 49, pp. 816–828, 2011.
6. K. Bashar and M. Murshed, "Texture Based Vein Biometrics for Human Identification: A Comparative Study," *2018 IEEE 42nd Annual Computer Software and Applications Conference (COMPSAC)*, 2018, pp. 571–576, 2018
7. I. Boussaïd, J. Lepagnot, P. Siarry, "A survey on optimization metaheuristics," Inf Sci, vol. 237, pp. 82–117, 2013.
8. H. Ishibuchi, T. Yoshida and T. Murata, "Balance between genetic search and local search in memetic algorithms for multiobjective permutation flowshop scheduling," IEEE Trans Evol Comput, vol. 7, pp. 204–223, 2003.
9. D.H. Wolpert and W.G. Macready, "No free lunch theorems for optimization," IEEE Trans Evol Comput, vol. 1, pp. 67–82, 1997.
10. H. Mohmmadzadeh and F.S Gharehchopogh, "An efficient binary chaotic symbiotic organisms search algorithm approaches for feature selection problems," J Supercomputing, vol. 77, pp. 9102–9144, 2021.
11. M.Y. Cheng and D Prayogo, "Symbiotic Organisms Search: A new metaheuristic optimization algorithm," Computers & Structures, vol 139, pp. 98–112, 2014.
12. X. Xi, G. Yang, Y Yin and X. Meng, "Finger Vein Recognition with Personalized Feature Selection", Sensors, vol. 13, pp. 11243–11259, 2013.
13. J. Wu, and C. Liu, "Finger-vein pattern identification using SVM and neural network technique," Expert Syst. Appl, Vol. 38, pp. 14284–14289, 2011.
14. B. A. Rosdi, C. W. Shing and S.A. Suandi, "Finger vein recognition using local line binary pattern," Sensors, vol. 11, pp. 11357–11371, 2011.
15. X. Zhang and W. Wang, "Finger vein recognition method based on GLCM-HOG and SVM," 2020 IEEE 3rd International Conference on Information Systems and Computer Aided Education (ICISCAE), pp. 698–701, 2020.
16. L. Yang, G. Yang, X. Xi, K. Su, Q. Chen and Y. Yin, "Finger Vein Code: From Indexing to Matching," IEEE Transactions on Information Forensics and Security, vol. 14, pp. 1210–1223, May 2019.
17. C. Kauba, E. Piciucco, E. Maiorana, P. Campisi and A. Uhl, "Advanced Variants of Feature Level Fusion for Finger Vein Recognition," 2016 International Conference of the Biometrics Special Interest Group (BIOSIG), pp. 1–7, 2016.
18. J. Yang and X. Zhang, "Feature-level fusion of global and local features for finger-vein recognition," IEEE 10th INTERNATIONAL CONFERENCE ON SIGNAL PROCESSING PROCEEDINGS, pp. 1702–1705, 2010.
19. Y. Lu, S. Yoon and D. S. Park, "Finger Vein Recognition based on Matching Score-Level Fusion of Gabor Features," The Journal of Korean Institute of Communications and Information Sciences, vol. 3, pp. 174–182, 2013.
20. Database USM: http://drfendi.com/fv_usm_database/
21. http://www.sz.tsinghua.edu.cn/labs/vipl/thu-fvfdt.html
22. Z. Liu and S. Song, "An embedded real-time finger-vein recognition system for mobile devices," IEEE Trans. Consum. Electron, vol. 58, 2012.
23. G. Gavisiddappa, S. Mahadevappa and C. Patil, "Multimodal Biometric Authentication System Using Modified ReliefF Feature Selection and Multi Support Vector Machine," International Journal of Intelligent Engineering and Systems, vol. 13, pp. 1–12, 2020.
24. R. M. Haralick, "Statistical and structural approach to texture," Proceeding of the IEEE, vol. 67, pp. 786–804, 1979.
25. R.M. Haralick, K. Shanmugam and I. Dinstein, "Textural Features for Image Classification," IEEE Trans. on Systems, Man and Cybernetics, Vol. SMC-3, pp. 610–621, 1973.

26. D. H. Tran, M. Y. Cheng and D. Prayogo, "A Novel Multiple Objective Symbiotic Organisms Search (MOSOS) for Time–Cost–Labor Utilization Tradeoff Problem". Knowledge-Based Systems, 94, 132–145, 2016.
27. A. E. Ezugwu and D. Prayogo, "Symbiotic organisms search algorithm: Theory, recent advances and applications," Expert Syst. Appl., vol. 119, pp. 184–209, Apr. 2019.
28. A. E. Ezugwu, O. J. Adeleke and S. Viriri, "Symbiotic organisms search algorithm for the unrelated parallel machines scheduling with sequence-dependent setup times," PloS one, vol. 13, pp. 1–23, 2018.

Part III
Big Algorithms for Software Development

Streamlining KYC Processes for Banks with Centralized Information Repository Using Design Thinking

T. R. Lekhaa and N. Ramya Devi

1 Introduction

Know Your Customer (KYC) is a critical process in the banking industry to identify and verify the identity of customers and assess potential risks of financial crimes such as money laundering, terrorism financing, and fraud. Traditional KYC processes can be time-consuming and costly and often result in a poor customer experience. However, the use of a centralized information repository that stores customer data can streamline the KYC process for banks [1].

Design thinking is a human-centered approach to problem-solving that can help develop effective solutions by understanding the needs and expectations of users. In this chapter, we propose using design thinking to create a centralized information repository for KYC processes that meets the needs of both customers and banks. This chapter will first identify the limitations of traditional KYC processes and the potential benefits of a centralized information repository. We will then discuss the design thinking approach and its application to the development of a centralized information repository for KYC processes. The chapter will provide a step-by-step process of using design thinking, including problem understanding, problem definition, ideation, prototyping, and testing [2].

The proposed solution can help banks to improve the efficiency and accuracy of KYC processes, reduce costs, and enhance customer experience. Additionally, the solution can facilitate compliance with regulatory requirements related to KYC. We will also discuss the potential regulatory benefits of using a centralized information

T. R. Lekhaa (✉) · N. Ramya Devi
Department of Information Technology, SNS College of Engineering, Coimbatore, Tamil Nadu, India
e-mail: lekhaa.tr.it@snsce.ac.in; ramya.n.it@snsce.ac.in

A. Haldorai et al. (eds.), *6th EAI International Conference on Big Data Innovation for Sustainable Cognitive Computing*, EAI/Springer Innovations in Communication and Computing, https://doi.org/10.1007/978-3-031-54696-9_9

repository for KYC processes [3]. This chapter highlights the potential benefits of using design thinking to create a centralized information repository for KYC processes and encourages further research and development in this area.

1.1 Objective

The objective of this chapter is to propose a solution to streamline the KYC processes for banks using a centralized information repository and to demonstrate the application of design thinking in the development of this solution. Specifically, the chapter aims to:

1. Identify the limitations of traditional KYC processes and the potential benefits of a centralized information repository.
2. Apply the design thinking approach to develop a centralized information repository that meets the needs of both customers and banks.
3. Outline the step-by-step process of using design thinking in the development of the solution, including problem understanding, problem definition, ideation, prototyping, and testing.
4. Demonstrate how the proposed solution can improve the efficiency and accuracy of KYC processes, reduce costs, and enhance customer experience.
5. Highlight the potential regulatory benefits of using a centralized information repository for KYC processes.
6. Encourage further research and development in the use of design thinking for creating innovative solutions in the banking industry.

1.2 Scope

The scope of this chapter is to propose a solution for streamlining KYC processes for banks using a centralized information repository and to demonstrate the application of design thinking in the development of this solution. The chapter will cover the following topics:

1. The limitations of traditional KYC processes and the potential benefits of a centralized information repository.
2. The design thinking approach and how it can be applied to develop a centralized information repository for KYC processes.
3. The step-by-step process of using design thinking, including problem understanding, problem definition, ideation, prototyping, and testing.
4. The proposed solution and its potential to improve the efficiency and accuracy of KYC processes, reduce costs, and enhance customer experience.
5. The potential regulatory benefits of using a centralized information repository for KYC processes.

6. The limitations and challenges of implementing the proposed solution.
7. The future research directions and opportunities for using design thinking to create innovative solutions in the banking industry.

The chapter will focus on the application of design thinking in the development of a centralized information repository for KYC processes and will not cover other aspects of banking operations or compliance.

1.3 Problem Definition

KYC Connect is a Software as a Service (SaaS) tool designed to streamline the Know Your Customer (KYC) processes for banks. The tool automates the collection, management, and analysis of customer information by providing a centralized repository of data. KYC Connect allows banks to improve compliance, reduce operational costs, and enhance customer experience by simplifying the KYC process [4].

KYC Connect's central repository of information ensures that all customer data is up-to-date, accurate, and easily accessible. The tool leverages advanced analytics and machine learning algorithms to detect any suspicious activity and alert the bank's compliance team in real-time. KYC Connect is customizable, allowing banks to configure the tool to match their specific KYC requirements [5].

With KYC Connect, banks can significantly reduce the time and resources required to conduct KYC due diligence. The tool automates many of the manual and time-consuming tasks associated with KYC, such as document verification, identity checks, and risk assessment. KYC Connect ensures that banks can comply with regulatory requirements and meet their anti-money laundering (AML) obligations in a timely and efficient manner.

Overall, KYC Connect offers banks a cost-effective and scalable solution to improve their KYC processes. By providing a centralized repository of information, the tool helps banks to streamline their operations, improve compliance, and enhance customer experience.

1.4 Empathy

We understand that banks face several challenges when it comes to conducting the Know Your Customer (KYC) process. These challenges include lengthy and time-consuming manual processes, managing large volumes of data, and navigating complex regulatory requirements. In addition, banks must ensure that they are providing a positive customer experience while also meeting compliance obligations [6].

At KYC Connect, we empathize with these challenges and are committed to helping banks streamline their KYC processes. Our SaaS tool automates many of the manual and time-consuming tasks associated with KYC, such as document verification, identity checks, and risk assessment. Our platform provides banks with a centralized repository of information, ensuring that all customer data is up-to-date, accurate, and easily accessible. Our tool also uses advanced analytics and machine learning algorithms to detect any suspicious activity, alerting the bank's compliance team in real-time. This approach ensures that banks can comply with regulatory requirements and meet their anti-money laundering (AML) obligations in a timely and efficient manner.

We recognize that each bank has unique KYC requirements and that a one-size-fits-all solution is not effective. Our tool is customizable, allowing banks to configure the tool to match their specific KYC requirements [7]. This approach ensures that banks can tailor their KYC processes to meet their specific needs and provide a positive customer experience.

At KYC Connect, we are committed to helping banks improve their KYC processes and provide a better customer experience. Our tool offers banks a cost-effective and scalable solution that simplifies the KYC process and ensures compliance with regulatory requirements.

Customer Empathy Map

- Say: "I want to open a bank account quickly and easily."
- Think: "I hope this bank doesn't ask for too much personal information."
- Feel: "I'm worried about my personal information being shared or stolen."
- Do: "I might avoid opening a bank account if the process is too complicated or invasive."

1.5 Bank Empathy Map

- Say: "We need to comply with KYC regulations while maintaining efficient operations."
- Think: "We have to collect a lot of information from customers, but we also need to ensure their privacy."
- Feel: "We want to provide a good customer experience while minimizing costs and risks."
- Do: "We might have to invest in technology and automation to streamline KYC processes."

2 Related Works

Several works and papers have been published on the topic of streamlining KYC processes for banks using technology and innovation. Some relevant works and papers include:

1. "Digital KYC: A Transformational Tool for Customer Onboarding" by Deloitte. This paper discusses the potential benefits of digital KYC solutions for banks, including improved efficiency, reduced costs, and enhanced customer experience.
2. "KYC Optimization: Identifying the Right Solutions" by Accenture. This paper explores the challenges and opportunities of KYC optimization and provides recommendations for banks to improve their KYC processes.
3. "KYC Utilities: Transforming the Customer Onboarding Experience" by KPMG. This paper discusses the potential of KYC utilities to streamline the KYC process for banks and improve the customer onboarding experience.
4. "A Design Thinking Approach to KYC Process Innovation" by Capgemini. This paper proposes using design thinking to create innovative solutions for KYC processes and provides a case study of its application in a banking context.
5. "Centralizing KYC Information: The Future of Compliance" by Thomson Reuters. This paper discusses the potential benefits and challenges of using a centralized information repository for KYC processes and provides recommendations for its implementation.

These works and papers provide valuable insights into the challenges and opportunities of streamlining KYC processes for banks using technology and innovation. They also highlight the importance of considering the needs and expectations of both customers and banks in the development of solutions.

3 Existing Systems

The existing KYC system in banks typically involves a manual and time-consuming process of collecting customer information and verifying their identity. Banks rely on various sources of information, including government-issued identification documents, proof of address, and employment information, to verify a customer's identity and assess the risks associated with them.

This process can be challenging and expensive for banks, as they often need to allocate significant resources to collecting and verifying customer information. Additionally, the traditional KYC process can result in a poor customer experience, as customers may have to provide the same information repeatedly or wait for extended periods for their accounts to be opened.

To overcome these challenges, banks have started adopting technology-based solutions to streamline KYC processes. For example, some banks have implemented digital KYC solutions that allow customers to submit their identity documents and other information electronically. Other banks have adopted biometric technologies, such as facial recognition and fingerprint scanning, to verify customer identity quickly and securely.

Despite these innovations, the KYC process remains complex and challenging for banks. One significant challenge is the lack of a centralized information repository that stores customer information, making it challenging to access and verify information quickly and efficiently. Additionally, the lack of standardization and coordination between different banks can make the KYC process more complex and time-consuming for customers.

Therefore, there is a need for a centralized information repository that streamlines KYC processes for banks and provides a better customer experience. In the following sections, we will discuss how design thinking can be used to develop a centralized information repository that meets the needs and expectations of both banks and customers.

4 Proposed System

The proposed system aims to streamline KYC processes for banks by developing a centralized information repository that stores customer information securely and efficiently. The system will use design thinking to create a user-centric solution that meets the needs and expectations of both banks and customers.

The proposed system will have the following features:

1. Centralized Information Repository: The system will have a centralized database that stores customer information securely and efficiently. The database will be accessible by authorized personnel at different banks, reducing the need for duplicate requests for customer information.
2. Customer Information Management: The system will allow customers to manage their information securely and conveniently. Customers will be able to update their personal information, upload identification documents, and provide consent for data sharing.
3. Verification and Compliance: The system will use automated processes to verify customer information and assess the risks associated with them. The system will comply with regulatory requirements, including Anti-Money Laundering (AML) and Know Your Customer (KYC) regulations.
4. Collaboration and Standardization: The system will enable collaboration between different banks and promote standardization in the KYC process. This will reduce the time and resources needed to collect and verify customer information, improving the overall customer experience.

Design thinking will be used to develop the proposed system. This approach will involve understanding the needs and expectations of different stakeholders, including banks, customers, and regulatory bodies. Prototyping and testing will be used to refine the system and ensure that it meets the needs of all stakeholders.

The proposed system will improve the KYC process for banks and provide a better customer experience. By centralizing customer information, the system will reduce the time and resources needed to collect and verify customer information. Additionally, customers will have more control over their information, and the system will comply with regulatory requirements, reducing the risk of non-compliance penalties.

4.1 Ideate

To streamline KYC processes for banks, a centralized information repository could be designed using design thinking. This repository would store all the necessary information about customers and would be accessible to all bank staff, making the KYC process faster and more efficient.

Ideas for this centralized information repository could include:

1. Customer Self-service: Allow customers to submit their own information online or through an app, reducing the need for bank staff to gather information manually.
2. Digital Verification: Use digital verification methods such as facial recognition or biometrics to authenticate customer identities quickly and accurately.
3. AI-Powered Screening: Implement AI-powered screening to flag high-risk customers and automatically escalate them for further review.
4. Intuitive Dashboard: Design an intuitive dashboard that provides a comprehensive view of customer information, making it easy for bank staff to quickly access and review relevant data.
5. Collaboration Tools: Provide collaboration tools that allow bank staff to work together seamlessly and share information as needed, streamlining the KYC process.
6. Analytics and Reporting: Incorporate analytics and reporting tools that enable bank staff to track KYC completion rates, identify bottlenecks, and continuously improve the process.

By using design thinking to create a centralized information repository, banks can streamline KYC processes, reduce costs, and improve customer satisfaction.

4.2 System Architecture

The system architecture for the proposed centralized information repository for KYC processes in banks using design thinking will comprise the following components:

1. Front-End Interface: This component will allow customers to access and manage their information securely and conveniently. Customers will be able to update their personal information, upload identification documents, and provide consent for data sharing. The front-end interface will be user-friendly and accessible from different devices, including mobile phones and desktops.
2. Centralized Information Repository: This component will store customer information securely and efficiently. The database will be accessible by authorized personnel at different banks, reducing the need for duplicate requests for customer information. The information repository will be compliant with regulatory requirements, including AML and KYC regulations.

3. Data Analytics and Verification: This component will use data analytics and verification processes to ensure the accuracy and completeness of customer information. The system will use automated processes to verify customer information and assess the risks associated with them. The data analytics and verification component will be compliant with regulatory requirements and best practices.
4. Collaboration and Standardization: This component will enable collaboration between different banks and promote standardization in the KYC process. The component will allow different banks to access customer information securely and efficiently, reducing the time and resources needed to collect and verify customer information.
5. Security and Compliance: This component will ensure that the system is secure and compliant with regulatory requirements, including data privacy and protection regulations. The component will use advanced security measures, such as encryption, to protect customer information.

The proposed system architecture will be designed to be scalable and adaptable to different banks' needs and requirements. The system will be designed to integrate with existing systems and technologies, reducing the need for banks to invest in new infrastructure.

Overall, the proposed system architecture will provide a centralized and secure repository for customer information, making the KYC process more efficient, convenient, and compliant with regulatory requirements (Fig. 1).

5 Results and Discussions

However, the implementation of a centralized information repository for KYC processes using design thinking is expected to result in the following benefits:

1. Improved Efficiency: The proposed system will streamline the KYC process, reducing the time and resources required to collect and verify customer information. This will improve efficiency and reduce costs for banks.
2. Enhanced Customer Experience: The front-end interface will provide customers with a convenient and secure way to manage their information. Customers will be able to update their personal information and provide consent for data sharing, reducing the need for repetitive requests for customer information.
3. Better Risk Management: The data analytics and verification component will use automated processes to verify customer information and assess the risks associated with them. This will enable banks to better manage risk and comply with regulatory requirements.
4. Increased Collaboration: The collaboration and standardization component will enable different banks to access customer information securely and efficiently, reducing the time and resources needed to collect and verify customer information. This will promote collaboration between banks and reduce duplication of efforts.

5. Improved Compliance: The security and compliance component will ensure that the system is secure and compliant with regulatory requirements, including data privacy and protection regulations. This will reduce the risk of non-compliance and associated penalties (Fig. 2).

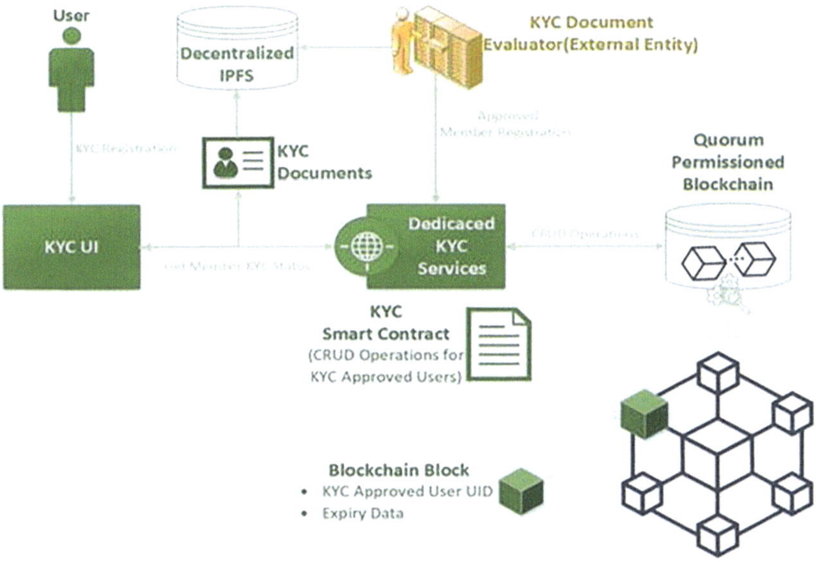

Fig. 1 KYC process

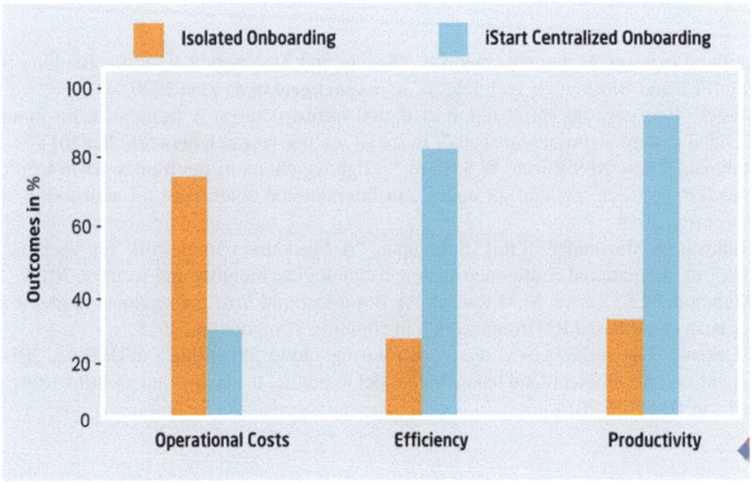

Fig. 2 Enhanced customer experience

Overall, the implementation of a centralized information repository for KYC processes using design thinking is expected to result in significant benefits for banks and their customers.

6 Conclusion

In conclusion, the proposed system of a centralized information repository for KYC processes in banks using design thinking can significantly improve the efficiency, security, and compliance of the KYC process. By leveraging design thinking principles, the system can be designed to meet the needs and requirements of different stakeholders, including customers, banks, and regulators.

The proposed system architecture, which includes a front-end interface, centralized information repository, data analytics and verification, collaboration and standardization, and security and compliance components, provides a comprehensive solution for streamlining the KYC process.

While the implementation of the proposed system may require significant investments in terms of time, resources, and technology, the potential benefits in terms of improved efficiency, enhanced customer experience, better risk management, increased collaboration, and improved compliance are substantial.

Overall, the proposed system of a centralized information repository for KYC processes in banks using design thinking represents a significant opportunity for banks to improve their KYC processes and stay competitive in an increasingly digital and regulatory environment.

References

1. Abdulla al mamun, Mohamed abuyusaf, "Secure and Transparent KYC for Banking System Using IPFS and Blockchain Technology" in researchgate.in on june 2020.
2. F.Glaster "Pervasive decentralization of digital infrastructures: A framework for blockchain embedded system and usecseanalysis", in social science research network, Jan 2017
3. A.Rahman, S.Roy, M.S.Kaiser, M.S.Islam, "A lightweight multi-tier framework to secure communication between low-end iot nodes", in International conference on networking, system and security, 2018
4. M.Raikwar, S.Mazumdar, S.Raj, S.Segupta, "A blockchain framework for insurance processes", in international conference on new technologies, mobility and security, 2018.
5. M.Mahmud, M.S. Kaiser, M.M.Rahaan "A Brain-inspired trust management model to assure security in cloud based IOT framework", in cognitive computation, 2018.
6. T.R.Lekhaa, "Networked visual servo control using cloud computing", in IJMTES, 2014
7. T.R.Lekhaa, "An efficient load balancing model to predict the best cloud partition using public cloud", in IJMTES, 2014.

Transforming Agriculture Through Internet of Things

Praveen M. Kulkarni, Padma S. Dandannavar, and Prayag Gokhale

1 Introduction

The global population is expected to be 9 billion by the year 2050, and it is expected that the world will face a severe shortage for agricultural products. There is a need for development of long-term measures for producing quality agricultural products and also reduce food insecurity across the globe. Conventional agricultural techniques are not supportive to meet the increasing the demand for agricultural products. Hence, to manage this situation, technology has been supportive for increasing the agricultural produce, technology like the Internet of Things (IoT) has been supportive for agricultural sector. Therefore, combination of technology and traditional agricultural practices would support in transforming the agricultural sector [1].

The agricultural sector can undergo a significant transformation, enabling farmers to achieve higher yields, optimize resource utilization, and mitigate environmental impacts. This chapter provides details with regard to the use, assistances, challenges, and scenarios for Internet of Things technology for agricultural [2].

P. M. Kulkarni (✉)
KLS, Institute of Management Education and Research, Belagavi, Karnataka, India

P. S. Dandannavar
Department of Information Science, KLS, Gogte Institute of Technology, Belagavi, Karnataka, India
e-mail: padmad@git.edu

P. Gokhale
KLE Technological University Dr. M S Sheshgiri College of Engineering and Technology, Belagavi, Karnataka, India

A. Haldorai et al. (eds.), *6th EAI International Conference on Big Data Innovation for Sustainable Cognitive Computing*, EAI/Springer Innovations in Communication and Computing, https://doi.org/10.1007/978-3-031-54696-9_10

According to the economic survey of India in the year 2020–2021, the Indian agricultural sector has more than 50% of the people working, and they are supporting the country's gross domestic product (GDP) with 20.2%.

Application of latest technology Internet of Things is supported by sensors and devices; farmers can access real-time information and actionable insights, facilitating informed decision-making for improved crop management, livestock monitoring, and supply chain optimization [3].

One notable area where IoT demonstrates its potential is in smart irrigation management. Conventional irrigation practices often result in excessive liquid usage, leading to waste and inefficient crop growth. However, by leveraging IoT-enabled sensors, farmers can screen soil humidity and climate surroundings. Data empowers farmers with irrigation scheduling, delivering precise amounts of water to crops based on their specific needs. As a result, water wastage can be significantly reduced, leading to enhanced water conservation and improved crop productivity [4].

Moreover, IoT-based crop monitoring systems enable farmers to track various parameters such as moisture, nutrient, pest infestations. The monitoring allows for the early detection of crop diseases, pest outbreaks, and nutrient deficiencies, enabling proactive measures. Timely interventions can mitigate crop losses and reduce the dependence on pesticides and encouraging sustainable farming practices [5].

Livestock monitoring is another area where IoT holds promise in Indian agriculture. Through the use of IoT devices, farmers can remotely monitor key pointers such as body temperature and heart rate of livestock. This capability facilitates early detection of health issues and optimization of animal husbandry practices. Improved livestock monitoring enhances animal welfare, disease control, and overall herd management [6].

The assitance of this technology for agriculture have certain challenges, which must address widespread adoption. These challenges include limited connectivity in rural areas, cost, security, privacy, and the need for farmer education and awareness. This chapter explores these constraints and provides potential direction for fostering successful integration of IoT in Indian agriculture [7].

Future research should prioritize the development of scalable and cost-effective IoT solutions, addressing interoperability issues, and promoting knowledge transfer and capacity building among farmers.

The amalgamation of these technologies for farming presents an incredible opportunity for the Indian farming community to enhance productivity, sustainability, and profitability. This study is focused on comprehensive review related to present situation, challenges, and future prospects of IoT in Indian agriculture. By addressing these challenges and harnessing the potential of IoT, the Indian agricultural sector can undergo a transformative shift toward efficient and sustainable farming attainment.

2 Literature Review

2.1 Agriculture in India

The role of agriculture is important for the Indian economy, contributing significantly with an 18% gross domestic product (GDP) and engaging approximately 57% of the rural population. However, despite the overall increase in agricultural output in India, there has been a decline in the number of farmers, dropping from 71.9% in 1951 to 45.1% in 2011 [8].

According to the Economic Survey 2018, the proportion of agricultural workers in the total workforce is projected to further decrease to 25.7% by 2050. Rural farming families face numerous challenges, including rising cultivation costs, low per capita productivity, inadequate soil maintenance, and a shift toward non-farming or higher-paying occupations [9].

As India stands on the brink of a digital revolution, it is the perfect opportunity to link agricultural and wireless technology, establishing digital connectivity for farmers. This connectivity can bring numerous benefits to the agricultural sector, including improved access to information, enhanced decision-making, and increased productivity. By leveraging wireless technology and introducing digital connectivity, farmers can overcome some of the challenges they face and create a more sustainable and efficient agricultural system [10].

The Indian economy is supported by the agricultural sector, and over the years, there has been an increase in the contribution to economy by agriculture. As per the economic survey conducted in the year 2020–2021, agricultural sector is contributing about 20% to the gross domestic product (GDP) [11].

Contemporary agriculture has transformed and changed in the recent past, as there is a significant support from the technology, which collects information from devices and sensor boards [12]. This technology also offers the farmers with information related to seeds, fertilizers, soil information, which support in producing quality crops. There are also information portals which give the farmers information on the latest development in the farming [13].

2.2 Technological Advancement in Indian Agriculture

In recent years, the agricultural sector in India has experienced significant technological advancements, leading to a transformation in traditional farming practices. These technological interventions aim to enhance productivity, reduce input costs, minimize post-harvest losses, and improve market access for farmers. This chapter analyzes several key technologies that have made an impact on Indian agriculture.

Precision Farming: This technology is based on sensors, GIS, GPS, and IoT. The benefits of precision farming include improved crop management, optimized resource utilization, and increased productivity [14].

Drones: Unmanned aerial vehicles or drones have emerged as powerful tools for crop monitoring and management. Equipped with sensors and cameras, drones assist farmers in detecting crop diseases, monitoring plant health, and optimizing resource allocation [15].

Agricultural Biotechnology: Biotechnology plays a crucial role in improving crop traits and combating pests and diseases. Genetically modified crops, such as Bt cotton, have led to increased yields and reduced pesticide use in India [16].

Farm Mechanization: The introduction of farm machinery and equipment has revolutionized agricultural practices in India, leading to increased efficiency, reduced labor dependency, and enhanced productivity [17].

Weather Forecasting and Remote Monitoring: Access to accurate weather information and remote monitoring technologies empowers farmers to make informed decisions. Mobile phone-based weather information dissemination helps farmers plan irrigation and protect crops from adverse weather conditions [18].

Mobile Applications: Mobile applications have become valuable tools for farmers, providing real-time information on weather, market prices, and agricultural practices [19].

Soil Health Management: Effective soil health management practices, including soil testing and nutrient management, contribute to sustainable agriculture. Soil health assessment methods are crucial in optimizing fertilizer usage and maintaining soil fertility [20].

Irrigation Technologies: Water scarcity is a significant challenge in Indian agriculture. Micro-irrigation techniques such as sprinkler and drip irrigation have shown potential in conserving water, improving crop productivity, and enhancing livelihoods [21].

E-Marketplaces: Digital platforms and e-marketplaces facilitate direct farmer-buyer interactions, reducing market inefficiencies and ensuring fair prices for agricultural produce in India [22].

Technological advancements have brought about a revolution in the agricultural landscape of India, providing opportunities for increased productivity, sustainability, and empowerment of farmers. However, it is important to ensure that these technologies are accessible and affordable to all farmers, particularly small-scale and marginalized ones. Continuous study, investment, and procedural backing fully harness the technology in Indian agriculture.

2.3 Internet of Things (IoT) in Indian Agriculture

The term "Internet of Things" (IoT) was coined by Kevin Ashton, co-founder and executive director of the Massachusetts Institute of Technology (MIT) Auto-ID Laboratory, during a presentation at Procter and Gamble (P&G) where radio-frequency identification (RFID) tags for supply chain management, enabling easier monitoring of item location and stock levels was familiarized [23].

There was an introduction of Internet Digital DIOS refrigerator in 2000, which utilized RFID technology and connected to the internet. Although the refrigerator did not achieve significant sales due to its perceived high cost, it paved the way for further internet-connected household appliances [24].

The Cisco Internet Business Solutions Group (IBSG) states that while the term "IoT" was coined in 1999, the actual concept emerged between 2008 and 2009 when the number of connected devices surpassed the global human population.

The growth of smartphones, tablet PCs, and smart gadgets led to 12.5 billion connected devices in 2010, with the number of devices per person exceeding 1 for the first time [25].

In the agricultural sector of India, the Internet of Things (IoT) has made remarkable progress, offering innovative solutions to address various challenges faced by farmers. IoT refers to a network of interconnected devices and sensors that collect and exchange data, enabling real-time monitoring, analysis, and decision-making [26].

Key applications of IoT in Indian agriculture include the following:

Smart Farming: IoT enables the implementation of smart farming practices by incorporating sensor-based monitoring systems into agricultural operations. These sensors collect information on ecological aspects such as soil humidity and light intensity. Farmers can remotely access this data to make informed decisions about irrigation, fertilization, and pest management [27].

Precision Agriculture: By integrating IoT devices and data analytics, precision agriculture techniques have gained traction in India. Farmers can monitor and manage their fields at a detailed level, optimizing resource utilization and increasing crop productivity. IoT-based precision agriculture provides insights into crop health, nutrient deficiencies, and disease outbreaks, enabling timely interventions [28].

Livestock Monitoring: IoT solutions extend to livestock management, where wearable sensors and tracking devices monitor animal health, location, and behavior. This technology aids in early disease detection, estrus cycle monitoring, and optimizing feed management. Farmers can track vital parameters remotely and receive alerts for any abnormalities [29].

Automated Irrigation Systems: IoT-powered automated irrigation systems facilitate efficient water management in agriculture. Sensors measure soil moisture levels, weather conditions, and crop water requirements, enabling precise irrigation scheduling. This technology minimizes water wastage and reduces labor associated with manual irrigation methods [30].

Supply Chain Management: IoT applications enable seamless integration throughout the agricultural supply chain, from farm to market. Real-time tracking of produce during transportation and storage ensures better quality control and reduces post-harvest losses. IoT-enabled traceability systems enhance food safety and enable effective recalls if necessary [31].

Market Access and Farming Advice: IoT applications provide with access of market data, prices, and demand trends. This allows them for choice crop harvesting and marketing approaches. IoT-based advisory services offer personalized recommendations and farming guidance based on local conditions, helping farmers optimize their operations [32].

Weather Monitoring and Forecasting: IoT devices are utilized for real-time weather monitoring, providing accurate forecasts to farmers. This information assists in planning farm activities, predicting pest and disease outbreaks, and mitigating weather-related risks. Accessible weather data helps farmers make timely decisions about planting, irrigation, and crop protection [33].

Farm Equipment Monitoring: IoT solutions enable remote monitoring and diagnostics of agricultural machinery, ensuring timely maintenance and reducing downtime. Sensors embedded in equipment provide data on fuel levels, engine performance, and operating parameters, allowing farmers to optimize machinery usage and increase operational efficiency [34].

Insurance and Risk Management: IoT-based technologies offer data-driven insights for agricultural insurance and risk management. Real-time monitoring of crop and weather conditions helps insurance companies accurately assess risks and determine premiums. This assists farmers in securing appropriate coverage and receiving timely compensation in case of crop losses [35].

The application of this technology for agriculture would enhance productivity, optimize resource utilization, reduce costs, and improve farmer incomes. It is essential to address challenges such as affordability, accessibility, and digital literacy across all segments of the farming community.

2.4 *Challenges of Internet of Things (IoT) Implementation in Indian Agriculture*

This technology has garnered significant support across various industries, including agriculture. In India, where agriculture significantly contributes to the economy, integrating IoT in this sector holds the potential for heightened efficiency and improved yields. Nonetheless, several obstacles impede the successful adoption of IoT in agriculture [36].

The study conducted by two organizations, namely, NASSCOM and CISCO to understand the IoT-enabled agriculture in India. The study results have indicated that there is a larger gap with regard to the application of this technology in the Indian agriculture. However, the study also highlighted that start-ups in the Indian agriculture sector are taking major steps to improve the agricultural output through this technology. But the major challenges still remain in the bridging the gap with the global adoption of this technology and adoption in the Indian scenario [37–39].

The primary hurdles in implementing advanced IoT solutions extend beyond individual sensors, focusing on difficulties in expanding the technology and showcasing its tangible, repeatable effects. These obstacles primarily stem from farmers' reluctance to embrace change and the constrained return on investment for solution providers [40].

3 Research Methods

The study examines the role of IoT technology in agriculture. The study has collected data from the three respondents, namely, (a) farmer, (b) farm equipment resellers, and (c) agriculture produce reseller. The study has applied the partial least square method for analyses of the data. The data was collected from 30 members from each group with a total of 90 respondents. The profile of the respondents is presented in Table 1.

3.1 *Variables for the Study*

The study included the variables derived from the systematic literature review and through the research gap. The details with regard to the study variables are presented in Table 2.

Table 1 Profile of the respondents for the study

	Male		Female		Total	
Gender	*N*	%	*N*	%	*N*	%
Farmers	28	37.84	5.00	31.25	33.00	36.67
Farm equipment resellers	17	22.97	8.00	50.00	25.00	27.78
Agriculture produce resellers	29	39.19	3.00	18.75	32.00	35.56
Total	**74**	**100.00**	**16**	**100.00**	**90**	**100.00**
Age	**N**	%				
25–30	47	52.22				
30–35	12	13.33				
35–40	11	12.22				
40–45	9	10.00				
45 and above	11	12.22				
Total	**90**	**100.00**				

Table 2 Study variables

Sr. no.	Variables	Source
1	Awareness of IoT in agriculture	[41, 42]
2	Use of IoT in agriculture	[43, 44]
3	Advantage of IoT in agriculture	[45, 46]
4	IoT needs more development in agriculture	[47, 48]

Table 3 Reliability analysis

Construct	Cronbach's alpha(α)
1. Awareness of IoT in agriculture	0.862
2. Use of IoT in agriculture	0.891
3. Advantage of IoT in agriculture	0.951
4. IoT needs more development in agriculture	0.865

4 Results

The results of the study are presented in Tables 3, 4 and 5. The results with regard to the reliability of the scale have shown a Cronbach alpha in Table 3 is 0.862 for awareness of IoT in agriculture, 0.891 for use of IoT in agriculture, 0.951 for advantage of IoT in agriculture, and 0.865 for scope for development of IoT technology in agriculture. Overall, the reliability scores have shown greater than (>) 0.65, which is acceptable for the research.

The results with regard to the convergent validity and scores from the analysis of average variance extracted (AVE) shown in Table 4 indicate that 0.756 for awareness of IoT in agriculture, 0.863 for use of IoT in agriculture, 0.762 for advantage of IoT in agriculture, and 0.951 for scope for development of IoT technology in agriculture. Overall, the convergent validity and scores from the analysis of average variance extracted (AVE) have shown greater than (>) 0.50, which indicate that the study variables have positive validity.

The results with regard to PLS analysis show that awareness of application of IoT in agriculture has increased among the study respondents with *P* value of 0.002, which is less than (<) 0.005 level of significance. This shows that awareness with regard to new technology in the domain of agriculture has increased among the users.

The results with regard to the use of Internet of Things (IoT) in agriculture have shown an increase with *P* value of 0.004, which is less than (<) 0.005 level of significance. This indicates that use of this technology in agriculture has increased, and the respondents have also indicated that Internet of Things (IoT) in agriculture is advantageous and hence the *P* value of 0.001, which is less than (<) 0.005 level of significance.

However, the results with regard to the technology development in the domain of Internet of Things (IoT) in agriculture need more development, and respondents

Table 4 Convergent validity

Construct	Average variance extracted (AVE)
1. Awareness of IoT in agriculture	0.756
2. Use of IoT in agriculture	0.863
3. Advantage of IoT in agriculture	0.762
4. IoT needs more development in agriculture	0.951

Table 5 PLS analysis

Effect	Beta	Total effect	*P* v alue
Respondents → Awareness of IoT in agriculture	0.0708	0.0708	0.002
Respondents → Use of IoT in agriculture	0.1875	0.1875	0.004
Respondents → Advantage of IoT in agriculture	0.1542	0.1542	0.001
Respondents → IoT needs more development in agriculture	0.1063	0.1063	0.311

have indicated the results with *P* value of 0.311, which is greater than (>) 0.005 level of significance.

5 Discussions

The study results have indicated a positive result with regard to the use of IoT and advantage of IoT in agriculture. However, there is still more development are needed, to specify the development IoT technologies can significantly enhance the efficiency, productivity, and sustainability of agricultural practices, which is of utmost importance for a country like India, where agriculture plays a vital role in the economy and employs a large portion of the population.

Here are some key reasons for the need for development in the domain of IoT in agriculture in India:

(a) IoT technologies enable precision agriculture, where farmers can analyze data from various sources, such as soil sensors, drones, and satellite imagery. This data-driven approach allows for precise and timely interventions, leading to better yields, reduced costs, and minimized environmental impact.
(b) A significant percentage of farmers in India are smallholders with limited access to modern agricultural practices and technology. IoT solutions can level the playing field by providing affordable and accessible tools to improve their farming practices.
(c) Developing the IoT in agriculture sector can stimulate economic growth and create new job opportunities, especially in technology-related fields, contributing to the overall development of the country.

(d) As other countries invest in smart agricultural technologies, India must remain competitive in the global agricultural market. Embracing IoT in agriculture can enhance India's agricultural competitiveness and help capture international market opportunities.
(e) IoT-driven farm automation is gradually transforming traditional farming practices in India. Automated systems for tasks such as planting, weeding, harvesting, and post-harvest handling are being developed and adopted to reduce labor dependency and increase productivity.

6 Conclusion

Hence, it is important to note that the adoption of IoT in agriculture can face challenges related to infrastructure, affordability, and digital literacy among farmers. However, with continued support from the government, private sector, and research institutions, IoT in agriculture is poised to play a transformative role in ensuring sustainable and efficient farming practices in India.

In conclusion, the development of IoT in agriculture in India is vital to address the challenges posed by a growing population, climate change, and resource constraints. By harnessing the power of IoT technologies, Indian agriculture can become more efficient, sustainable, and resilient, ensuring food security, economic growth, and improved livelihoods for millions of farmers.

References

1. Hubert, B., Rosegrant, M., Van Boekel, M. A., & Ortiz, R. (2010). The future of food: scenarios for 2050. *Crop Science*, *50*, S-33.
2. Ahmed, Nurzaman, Debashis De, and Iftekhar Hussain. "Internet of Things (IoT) for smart precision agriculture and farming in rural areas." IEEE Internet of Things Journal 5.6 (2018): 4890–4899.
3. Daoliang, L. I., & Chang, L. I. U. (2020). Recent advances and future outlook for artificial intelligence in aquaculture. Smart Agriculture, 2(3), 1.
4. Boursianis, A. D., Papadopoulou, M. S., Diamantoulakis, P., Liopa-Tsakalidi, A., Barouchas, P., Salahas, G., … & Goudos, S. K. (2022). Internet of things (IoT) and agricultural unmanned aerial vehicles (UAVs) in smart farming: A comprehensive review. Internet of Things, 18, 100187.
5. Abu, N. S., Bukhari, W. M., Ong, C. H., Kassim, A. M., Izzuddin, T. A., Sukhaimie, M. N., … & Rasid, A. F. A. (2022). Internet of things applications in precision agriculture: A review. Journal of Robotics and Control (JRC), 3(3), 338–347.
6. Kumar, K., Kumar, A., Kumar, N., Mohammed, M. A., Al-Waisy, A. S., Jaber, M. M., … & Al-Andoli, M. N. (2022). Dimensions of internet of things: Technological taxonomy architecture applications and open challenges—A systematic review. Wireless Communications and Mobile Computing, 2022.
7. Navarro, E., Costa, N., & Pereira, A. (2020). A systematic review of IoT solutions for smart farming. Sensors, 20(15), 4231.

8. Tiwari, A., & Jaga, P. K. (2012). Precision farming in India—A review. Outlook on Agriculture, 41(2), 139–143.
9. Sishodia, R. P., Ray, R. L., & Singh, S. K. (2020). Applications of remote sensing in precision agriculture: A review. Remote Sensing, 12(19), 3136.
10. Singh, R. L., & Mondal, S. (Eds.). (2017). Biotechnology for sustainable agriculture: emerging approaches and strategies.
11. Benos, L., Tsaopoulos, D., & Bochtis, D. (2020). A review on ergonomics in agriculture. part II: Mechanized operations. Applied Sciences, 10(10), 3484.
12. Yegbemey, R. N., & Egah, J. (2021). Reaching out to smallholder farmers in developing countries with climate services: A literature review of current information delivery channels. Climate Services, 23, 100253.
13. Laurett, R., Paço, A., & Mainardes, E. W. (2021). Sustainable development in agriculture and its antecedents, barriers and consequences—an exploratory study. Sustainable Production and Consumption, 27, 298–311.
14. Hussain, Z., Deng, L., Wang, X., Cui, R., & Liu, G. (2022). A review of farmland soil health assessment methods: Current status and a novel approach. Sustainability, 14(15), 9300.
15. Assefa, T., Jha, M., Reyes, M., Tilahun, S., & Worqlul, A. W. (2019). Experimental evaluation of conservation agriculture with drip irrigation for water productivity in sub-Saharan Africa. Water, 11(3), 530.
16. Srivastava, G. (2022). Antecedents of E-Marketing of Agriculture Products in This Digital Era: An Empirical Study. International Journal of Technology and Human Interaction (IJTHI), 18(7), 1–17.
17. Afridi, F., Bishnu, M., & Mahajan, K. (2023). Gender and mechanization: Evidence from Indian agriculture. American Journal of Agricultural Economics, 105(1), 52–75.
18. Ahire, D. B., Gond, D. V. J., & Ahire, N. L. (2022, February). IoT Based Real-Time Monitoring of Meteorological Data: A Review. In Proceedings of the 3rd International Conference on Contents, Computing & Communication (ICCCC-2022).
19. Shams, R. A., Shahin, M., Oliver, G., Perera, H., Whittle, J., Nurwidyantoro, A., & Hussain, W. (2023). Investigating end-users' values in agriculture mobile applications development: An empirical study on Bangladeshi female farmers. Journal of Systems and Software, 200, 111648.
20. Gutknecht, J., Journey, A., Peterson, H., Blair, H., & Cates, A. (2023). Cover crop management practices to promote soil health and climate adaptation: Grappling with varied success from farmer and researcher observations. Journal of Environmental Quality, 52(3), 448–464.
21. Serote, B., Mokgehle, S., Senyolo, G., du Plooy, C., Hlophe-Ginindza, S., Mpandeli, S., ... & Araya, H. (2023). Exploring the Barriers to the Adoption of Climate-Smart Irrigation Technologies for Sustainable Crop Productivity by Smallholder farmers: Evidence from South Africa. Agriculture, 13(2), 246.
22. Srivastava, P. K., Kumar, S., Tiwari, A., Goyal, D., & Mamodiya, U. (2023, June). Internet of thing uses in materialistic ameliorate farming through AI. In AIP Conference Proceedings (Vol. 2782, No. 1). AIP Publishing.
23. Saha, P., Kumar, V., Kathuria, S., Gehlot, A., Pachouri, V., & Duggal, A. S. (2023, May). Precision Agriculture Using Internet of Things and Wireless Sensor Networks. In 2023 International Conference on Disruptive Technologies (ICDT) (pp. 519–522). IEEE.
24. Chandana, M. S., Sridevi, C., Chowdary, R. A., & Likhitha, D. (2023, March). Automated Aquaponics Farming using Internet of Things (IoT). In 2023 Second International Conference on Electronics and Renewable Systems (ICEARS) (pp. 643–647). IEEE.
25. Polkowski, Z., Mishra, S. K., Mishra, B. K., Borah, S., & Mohanty, A. (2023). Impact of internet of things in social and agricultural domains in rural sector: a case study. International Journal of Cloud Computing, 12(1), 90–105.
26. Royal, N. S., Parre, W., Bandarupalli, S. C., Achary, M. R., Jadala, V. C., & Kavitha, M. (2023, January). Internet of Things (IoT) and Machine Learning based Optimized Smart Irrigation System. In 2023 5th International Conference on Smart Systems and Inventive Technology (ICSSIT) (pp. 874–878). IEEE.

27. Royal, N. S., Parre, W., Bandarupalli, S. C., Achary, M. R., Jadala, V. C., & Kavitha, M. (2023, January). Internet of Things (IoT) and Machine Learning based Optimized Smart Irrigation System. In 2023 5th International Conference on Smart Systems and Inventive Technology (ICSSIT) (pp. 874–878). IEEE.
28. Asha, K. N., Gahana, N., Gowda, S., Harsha, C. R., & Chandana, D. M. (2023). Agriculture Automation System using Machine Learning and Internet of Things. International Journal of Engineering and Management Research, 13(3), 163–167.
29. Kumar, G. S., Nagaraju, G., Rohith, D., & Vasudevarao, A. (2023). Design and Development of Smart Irrigation System Using Internet of Things (IoT)—A Case Study. Nature Environment and Pollution Technology, 22(1), 523–526.
30. Pillai, R., & Sivathanu, B. (2020). Adoption of internet of things (IoT) in the agriculture industry deploying the BRT framework. Benchmarking: An International Journal, 27(4), 1341–1368.
31. Sekaran, K., Meqdad, M. N., Kumar, P., Rajan, S., &Kadry, S. (2020). Smart agriculture management system using internet of things. TELKOMNIKA (Telecommunication Computing Electronics and Control), 18(3), 1275–1284.
32. Ayaz, M., Ammad-Uddin, M., Sharif, Z., Mansour, A., & Aggoune, E. H. M. (2019). Internet-of-Things (IoT)-based smart agriculture: Toward making the fields talk. IEEE access, 7, 129551–129583
33. Singh, G., Sharma, D., Goap, A., Sehgal, S., Shukla, A. K., & Kumar, S. (2019, October). Machine Learning based soil moisture prediction for Internet of Things based Smart Irrigation System. In 2019 5th International Conference on Signal Processing, Computing and Control (ISPCC) (pp. 175–180). IEEE.
34. Sinha, A., Shrivastava, G., & Kumar, P. (2019). Architecting user-centric internet of things for smart agriculture. Sustainable Computing: Informatics and Systems, 23, 88–102.
35. Madushanki, A. R., Halgamuge, M. N., Wirasagoda, W. S., & Syed, A. (2019). Adoption of the Internet of Things (IoT) in agriculture and smart farming towards urban greening: A review. International Journal of Advanced Computer Science and Applications, 10(4), 11–28.
36. Ayaz, M., Ammad-Uddin, M., Sharif, Z., Mansour, A., & Aggoune, E. H. M. (2019). Internet-of-Things (IoT)-based smart agriculture: Toward making the fields talk. IEEE access, 7, 129551–129583.
37. Jaiswal, S. P., Bhadoria, V. S., Agrawal, A., & Ahuja, H. (2019). Internet of Things (IoT) for smart agriculture and farming in developing nations. International Journal of Scientific & Technology Research, 8(12), 1049–1056.
38. Subeesh, A., & Mehta, C. R. (2021). Automation and digitization of agriculture using artificial intelligence and internet of things. Artificial Intelligence in Agriculture, 5, 278–291.
39. Khanna, A., & Kaur, S. (2019). Evolution of Internet of Things (IoT) and its significant impact in the field of Precision Agriculture. Computers and electronics in agriculture, 157, 218–231.
40. Yadav, S., Luthra, S., & Garg, D. (2021). Modelling Internet of things (IoT)-driven global sustainability in multi-tier agri-food supply chain under natural epidemic outbreaks. Environmental Science and Pollution Research, 28, 16633–16654.
41. Pillai, R., & Sivathanu, B. (2020). Adoption of internet of things (IoT) in the agriculture industry deploying the BRT framework. Benchmarking: An International Journal, 27(4), 1341–1368.
42. Yadav, E. P., Mittal, E. A., & Yadav, H. (2018, February). IoT: Challenges and issues in indian perspective. In 2018 3rd International Conference On Internet of Things: Smart Innovation and Usages (IoT-SIU) (pp. 1–5). IEEE.
43. Lakhwani, K., Gianey, H., Agarwal, N., & Gupta, S. (2019). Development of IoT for smart agriculture a review. In Emerging Trends in Expert Applications and Security: Proceedings of ICETEAS 2018 (pp. 425–432). Springer Singapore.
44. Patil, K. A., & Kale, N. R. (2016, December). A model for smart agriculture using IoT. In 2016 international conference on global trends in signal processing, information computing and communication (ICGTSPICC) (pp. 543–545). IEEE.
45. Kim, W. S., Lee, W. S., & Kim, Y. J. (2020). A review of the applications of the internet of things (IoT) for agricultural automation. Journal of Biosystems Engineering, 45, 385–400.

46. da Silveira, F., Lermen, F. H., & Amaral, F. G. (2021). An overview of agriculture 4.0 development: Systematic review of descriptions, technologies, barriers, advantages, and disadvantages. Computers and electronics in agriculture, 189, 106405.
47. Xu, J., Gu, B., & Tian, G. (2022). Review of agricultural IoT technology. Artificial Intelligence in Agriculture, 6, 10–22.
48. Ramachandran V, Ramalakshmi R, Kavin BP, Hussain I, Almaliki AH, Almaliki AA, Elnaggar AY, Hussein EE. Exploiting IoT and Its Enabled Technologies for Irrigation Needs in Agriculture. Water. 2022; 14(5):719.

Anti-theft System Using Structural Similarity

K. L. Abhishek, R. Geetanjali, S. Sushitha, and T. Shonit Aric

1 Introduction

In today's rapidly evolving world, ensuring the safety and security of our valuable possessions has become more crucial than ever. Traditional methods of preventing theft and unauthorized access, relying solely on human safety officers, have proven to be ineffective and unreliable. Our revolutionary framework introduces a comprehensive solution that not only detects intruders but also safeguards owners' property. By implementing our cutting-edge solution, we can significantly reduce the occurrence of theft. A pivotal aspect of our strategy for preventing potential property damage is swift response. Through the implementation of a robust anti-theft system equipped with advanced intruder detection capabilities, we can promptly identify unauthorized presence and immediately notify the owner or user using API alerts. Beyond theft prevention, our architecture extends to locating lost items. We have integrated a comprehensive lost and found database, simplifying the process of finding misplaced items. Utilizing our system's object detection techniques enhances the likelihood of successful retrieval. Central to our anti-theft technology is the potent computer vision approach of object detection. This methodology precisely locates and identifies items in images or video streams. Our system stands as an ideal choice for theft prevention and overall security enhancement due to its real-time monitoring and detection capabilities. Cameras strategically placed in designated areas continuously record videos to activate our architecture. Python- and OpenCV-based client-side algorithms are then applied to this video data. Any unauthorized entry triggers swift responses through motion and object detection

K. L. Abhishek (✉) · R. Geetanjali · S. Sushitha · T. S. Aric
M S Ramaiah Institute of Technology, Bangalore, India
e-mail: geetanjali.r@msrit.edu; sushithas@msrit.edu

A. Haldorai et al. (eds.), *6th EAI International Conference on Big Data Innovation for Sustainable Cognitive Computing*, EAI/Springer Innovations in Communication and Computing, https://doi.org/10.1007/978-3-031-54696-9_11

algorithms, instantly alerting users or owners via the connected API. Addressing the rising instances of theft and unauthorized access demands the development of an anti-theft system rooted in object detection techniques. Object detection, a technology widely used in areas such as self-driving cars, facial recognition, and medical imaging, proves its effectiveness and reliability. Harnessing this technology, our anti-theft system offers a dependable method to identify and prevent the theft of valuable items, ranging from vehicles to electronics.

In the ever-changing landscape of security challenges, the need for innovative and reliable anti-theft solutions has never been more critical. Our groundbreaking framework not only revolutionizes the way we address unauthorized access and theft but also redefines the concept of property protection. Departing from traditional reliance on human safety officers, our comprehensive solution incorporates cutting-edge technology to provide a robust defense against intruders and safeguard valuable possessions. The core of our strategy lies in the swift response to potential threats. With a state-of-the-art anti-theft system boasting advanced intruder detection capabilities, we can significantly reduce the occurrence of theft. Our system goes beyond mere identification, ensuring immediate action through API alerts that promptly notify the property owner or user of unauthorized presence. This quick response is a crucial element in minimizing property damage and enhancing overall security. Not limited to theft prevention, our architecture extends its functionality to locating lost items. We have seamlessly integrated a comprehensive lost and found database into our system, simplifying the process of recovering misplaced belongings.

The implementation of object detection techniques within our system further increases the likelihood of successful retrieval, adding an extra layer of utility and value to our innovative security solution. Central to the effectiveness of our anti-theft technology is the powerful computer vision approach of object detection. This cutting-edge methodology precisely locates and identifies items in images or video streams, making it an ideal choice for theft prevention and security enhancement. Cameras strategically placed in designated areas continuously record videos, activating our architecture to employ Python- and OpenCV-based client-side algorithms. These algorithms swiftly respond to any unauthorized entry, employing motion and object detection to instantly alert users or owners through the connected API. Addressing the rising instances of theft and unauthorized access requires a sophisticated anti-theft system rooted in advanced object detection techniques. Object detection, a technology widely employed in diverse fields such as self-driving cars, facial recognition, and medical imaging, has proven its effectiveness and reliability. By harnessing this technology, our anti-theft system offers a dependable method to identify and prevent the theft of valuable items, spanning from vehicles to electronics. Beyond its primary role in theft prevention, an object detection-based anti-theft system unlocks additional advantages.

The system's ability to track the movement of objects within the camera's field of vision enhances its utility in locating and recovering lost items. This innovative feature not only augments the usability of our intricate security architecture but also positions our solution as a comprehensive asset in managing the broader spectrum of security challenges. In conclusion, our revolutionary anti-theft framework represents a paradigm shift in the realm of security solutions. By combining cutting-edge object detection technology with swift response mechanisms and a comprehensive lost and found database, our system not only prevents theft but also contributes to the recovery of lost items. The versatility and effectiveness of our architecture make it a valuable asset in the ongoing battle against unauthorized access and property loss, setting a new standard for security in today's dynamic world.

Beyond its theft-prevention capabilities, an object detection-based anti-theft system offers advantages that extend further. The system can track the movement of an object within the camera's field of vision, enhancing the ability to locate and recover lost items. This innovation enhances the usability and clarity of our intricate security architecture. In conclusion, our sophisticated anti-theft system, leveraging the power of object detection techniques, presents a state-of-the-art solution to the ongoing challenges of theft and unauthorized access. Through computer vision, real-time monitoring, and rapid alert mechanisms, we provide a robust and dependable defense for safeguarding valuable possessions. Our framework offers a holistic security solution that not only deters theft but also traces and recovers lost items.

2 Literature Survey

The surge in theft incidents has created an unsafe atmosphere, inducing fear among individuals. Home security has emerged as a paramount concern, exacerbated by the prevalence of false alarms in existing intrusion detection systems. A potential remedy involves the integration of mobile phones and OpenCV to accurately identify intrusions while filtering out non-intrusive movements. This system archives footage locally and promptly notifies users of breaches through the Twilio API.

Current security systems typically require human officers or complex installations, leading to false alarms. While some systems use cameras to record video footage to identify intruders, manual inspection of the recordings takes time and may miss important details. While alarming systems alert owners when an intruder enters, radar-based systems use microwaves or radio frequency waves to identify intruders. Other systems use passive infrared motion detection, vibration sensors, or ultrasonic motion detection. These systems, however, lack effective post-processing techniques to precisely detect changes.

As a proof-of-concept for identifying people and locally storing video footage, the presented application is provided. It requires an Internet connection and an IP camera app and is designed for cellphones with a 5-megapixel camera. The program labels video clips with the appropriate file name, indicating the presence of humans, to produce good detection results. When an intrusion is discovered, SMS security notifications are further issued utilizing an SMS sending API. The program is appropriate for static environments where people move around, like conference rooms. The ability to discern between human movement and that of other moving objects, however, is limited [1]. In the paper "Smart Intruder Detection System with

OpenCV and Python," Bhanu Teja, Sandeep Kumar, and Dileep Reddy propose a system that utilizes OpenCV and a cell phone to reliably detect intruders while removing false alarms brought on by moving objects. When an incursion is found, the system notifies the user using the Twilio API and stores the video footage locally. The authors analyze the shortcomings of current security measures and highlight the benefits of their suggested remedy. In static environments with human activity, the system operates effectively. However, it might be less efficient in locations with bad illumination or poor camera quality. Its accuracy may also be impacted by environmental variables as temperature, humidity, and noise levels [2]. In the paper "Object Detection and Tracking in Wide Area Surveillance Using Thermal Imagery," Santosh Bhusal explores object detection and tracking techniques using thermal imagery. The author talks about the difficulties in finding and following things of interest and gives instances like counting deer using infrared imaging and spotting cars and people on surveillance footage. Various object detection and tracking techniques, including motion-based models and probabilistic models, are discussed. This document offers technical information that can be useful for researchers and industry experts, but it also makes some technical knowledge assumptions [3]. A study on "Object Detection" by Manjula S. and Dr. K. Lakshmi discusses the need for automated visual surveillance systems and the drawbacks of relying on human monitoring. The chapter focuses on the object detection's significance in visual surveillance, and the numerous methods for spotting moving things in video sequences are the main topics of this research. The chapter emphasizes the several uses of autonomous visual surveillance, including human identification, traffic monitoring, and spotting anomalous behavior [4]. The article "Spot the Difference Between Two Images Using Python" demonstrates how to use Python and the Pillow module to detect differences between two images. The structural similarity index (SSIM) of the images is calculated by the author using the ImageChops.difference() technique. The paper explains the application of the SSIM index beyond image processing and its usefulness as a similarity measure for comparing any two signals. The paper also briefly mentions video tracking and different levels of object tracking, including single object tracking and multiple object tracking. It also illustrates instances where object detection is used in surveillance. These papers provide insights into different aspects of object detection, including intrusion detection, thermal imagery, visual surveillance, and image comparison. Each paper analyzes the positives and negatives of the suggested solutions in detail and offers helpful knowledge to academics and industry experts in the subject of object detection. Its widespread adoption, robustness, and ease of integration make it a go to choice for computer vision tasks across various industries [5].

An automated comprehension and evaluation of data utilizing smart systems is now necessary due to the increase in picture and video data collections. In face identification, appearance recognition, human-computer interaction, and security systems, facial recognition technology is essential. It turns the features of the face into a distinctive face print that is kept in a database for verification. Deep learning systems identify objects by comparing digital photos to stored images. Applications for facial recognition include online exam proctoring, security, surveillance, and the consumer market. This study uses OpenCV, Face Recognition, and Python to create algorithms that address image quality issues in facial recognition systems already in use. The paper goes into the facial recognition technology's methodology, implementation, and uses. The process of doing a literature survey includes searching academic databases, reading research articles, analyzing results, and identifying research gaps that require more inquiry [6]. To solve the problem of single picture darkness, this research offers a novel technique called CvmD-net, which combines a reduction of color variation multi-branch decomposition network for retinex decomposition and enhancement. The two steps of the presented method are brightness optimization and image deconstruction. Reduction of color distortion and artefacts is made possible by the addition of an input constant feature prior method (ICFP) based on reflection constant features. To perform a variety of prediction tasks and reduce noise amplification, the decomposition network uses many subnetworks with various architectures. Overall, the CvmD-net method appears to be working well in resolving the problem of darkness in single photos [3]. For the purpose of detecting face masks on people, this project makes use of a variety of technologies, including OpenCV, a Caffe-based face detector, Keras, TensorFlow, and MobileNetV2. A total of 3835 photos make up the dataset used, with photographs from 1919 lacking masks and images from 1916 showing people wearing them [7, 8].

Using Keras and MobileNetV2, a base model is first produced. On top of the base model, a head model is then built, which has several layers, including a dropout layer and an activation function. The labeled dataset, which is divided into a training set (75% of the photos) and a test set (25% of the images) to assess model correctness, is used to train this combination model. Once trained, the model is loaded and used to analyze human face-containing pictures or video frames. By enlarging the input image or frame and detecting the blob, the basic face detector module can detect human faces. The trained model is then fed the cropped face to determine whether or not a mask is visible. Another model is additionally trained using OpenCV, specifically on pictures of people's faces linked to names and email addresses. The CV model recognizes faces in input images and requests the user for the person's name and email address, which are then recorded in a database. The first model's output is then checked against the database to determine who it belongs to. If a match is made, the person's name is drawn from a bounding box, and they are notified by email and SMS that they are not wearing masks. In every other case, the bounding box shows the words "Mask" if a mask is found, and "No Mask" if none is. Overall, this project combines different technologies to detect face masks on humans, provides identification, and sends notifications based on mask-wearing behavior [9, 10]. The loss function is made simpler in the modified Tiny YOLOv2 model for shuttlecock identification by using the width and height of the bounding box directly rather than their square roots. Additionally, the classification component of the loss function is eliminated for efficiency as there are only two classes (shuttlecock and backdrop). The model now performs and operates more effectively while detecting shuttlecocks as a result of these changes [11, 12]. This chapter introduces a brand-new cross-scale prediction module for precise content-adaptive single-reference prediction. The suggested module creates a reference feature pyramid that enables efficient motion compensation by utilizing characteristics at various scales. For straightforward motions, retaining context, and making accurate predictions, high-resolution reference characteristics are used. For complex motion scenarios, low-resolution r characteristics are used [13, 14].

In "LBPH Based Face Recognition at Low Resolution," Cao et al. and Ahmed et al. [15–17] discuss the difficulties in efficiently identifying individuals through video monitoring cameras, especially under conditions like blur, varying illumination, low resolution, and challenging lighting scenarios. In response, our proposed system excels even at a minimum resolution of 35px, demonstrating robust performance in identifying human faces across different angles, side poses, and during motion. We introduce the LR500 dataset for training and classification, and our paper pioneers the use of the Local Binary Patterns Histogram (LBPH) algorithm architecture for real-time, low-resolution human face recognition, providing a significant advancement in overcoming prevalent challenges.

These papers collectively illuminate diverse facets of object detection, spanning intrusion detection, thermal imagery, visual surveillance, and image comparison. Each work meticulously assesses the pros and cons of proposed solutions, providing valuable insights for both academics and industry practitioners in the realm of object detection. Given its broad adoption, robustness, and seamless integration, object detection remains a cornerstone in various industries' computer vision endeavors.

3 Methodology

3.1 Structural Similarity Index Measure

The structural similarity index measure (SSIM) is a technique for forecasting how well digital television, cinematic images, and other digital images and videos will be received by the public. SSIM is a tool for calculating how similar two photos are to one another. The SSIM index is based on its evaluation of image quality on a reference image that is uncompressed or distortion-free since it is a complete reference measure. The SSIM is a perception-based model that incorporates crucial perceptual phenomena, such as luminance and contrast masking terms, and views image degradation as a perceived change in structural information. The distinction between these methods and others is that they estimate absolute errors, unlike MSE or PSNR [18, 19]. The concept of structural information holds that pixels have high interdependencies, particularly when they are spatially close to one another. Important details regarding the organization of the items in the visual scene are carried by these dependencies. Luminance masking is a phenomenon where visual distortions (in this context) tend to be less noticeable in bright locations, whereas contrast masking is a phenomenon where distortions become less evident when there is significant activity or "texture" in the image [20].

3.2 Framework

Although specific models for evaluating perceptual quality might differ in detail, they generally adhere to a common structure. This framework comprises distinct stages:

1. *Pre-processing:* This initial phase involves operations aimed at rectifying known image distortions. Activities like scaling, alignment, color scheme alteration, pixel value conversion, low-pass filtering, and light adaptation simulation are part of this stage. The pre-processing phase rectifies image distortions through operations like scaling, alignment, color scheme alteration, pixel value conversion, low-pass filtering, and light adaptation simulation. Structural components include algorithms for resizing, geometric transformations, color space conversions, pixel value adjustments, filtering, and simulating diverse lighting conditions. These operations collectively enhance image quality and prepare it for subsequent analysis by addressing various distortions and optimizing features for effective image processing.
2. *CSF Filtering:* The Contrast Sensitivity Function (CSF) is incorporated in the proposed system, which accounts for the human visual system's sensitivity to different spatial and temporal frequencies. Quality assessment models often employ a CSF weighting stage, which might entail linear filtering or base-sensitivity normalization following channel decomposition. CSF filtering integrates the Contrast Sensitivity Function (CSF) to consider the human visual system's sensitivity across spatial and temporal frequencies. Quality assessment models utilize a CSF weighting stage, involving linear filtering or base-sensitivity normalization after channel decomposition. This structural component employs algorithms for precise CSF integration, channel decomposition, and subsequent operations that enhance the model's ability to accurately reflect human visual sensitivity, ensuring effective image quality assessment aligned with perceptual considerations.
3. *Channel Decomposition:* Image processing involves dividing images into sub-bands or channels that selectively capture orientation, spatial, and temporal frequencies. Structural components encompass techniques, ranging from intricate neural response-based transformations to simpler methods like discrete cosine transform (DCT) or wavelet transforms. These operations enable nuanced analysis, extracting essential features for subsequent processing or quality assessment, aligning with varying levels of complexity and computational efficiency.
4. *Error Normalization:* The error between decomposed reference and distorted signals is computed and then normalized. A masking model considers how certain visual components influence the visibility of surrounding elements, considering visibility thresholds. This could involve metrics like the just noticeable difference (JND) to quantify the error [21, 22].
5. *Error Pooling:* Normalized error signals are aggregated into a single value through error pooling. Often, a Minkowski norm is used, incorporating spatial extent and channel information. The pooling process might include varying exponents and spatially variable weighting to address different region importance.

This systematic framework systematically evaluates image quality by integrating pre-processing, CSF filtering, channel decomposition, error normalization, and error pooling stages. These steps facilitate quantifying image quality by accounting for human visual system characteristics and perceptual criteria.

The architecture diagram consists of a landing page and three models: face recognition, motion detection, and structural similarity. The face recognition model employs a trained face database for recognition. The motion detection model captures and records motion incidents, storing data for detection purposes. The structural similarity model compares images of lost items with reference photos in a dedicated database. The user interface on the landing page provides access to these models, enabling efficient management of recognized faces, motion incidents, and lost items.

The system features a user-friendly GUI allowing users to input IDs and names for face recognition training. The application comprises two modules: motion detection and spot mode. The former captures unauthorized activities, providing essential evidence. The latter rapidly identifies misplaced items, utilizing structural similarity concepts to aid in recovery. These modules enhance security measures and ensure prompt responses to unauthorized actions. By integrating face training, motion detection, and spot mode, the system adeptly identifies authorized individuals, records suspicious activities, and streamlines the process of locating lost items.

These steps enable the quantification of image quality by leveraging human visual system characteristics and perceptual criteria. In Fig. 1, the project's architectural diagram encompasses a landing page and three core models: face recognition, motion detection, and structural similarity. The face recognition model effectively employs a training face database to identify individuals. Meanwhile, the motion detection model records and captures instances of motion, storing this data in a dedicated database for subsequent motion detection analysis. In parallel, the structural similarity model compares images from a specialized reference photo database with images of misplaced items. The user interface on the landing page provides user-friendly access to these models. In the process of recognizing faces, the face recognition model relies on the training face database, while the motion detection model captures pertinent motion-related data. Utilizing the structural similarity model, comparisons are made between images to identify potential matches. Ensuring seamless system performance, essential data is meticulously maintained within the databases. The system is enhanced by an intuitive graphical user interface (GUI) that allows users to input identification and names for face recognition training. The motion detection module effectively captures activities related to unauthorized entry, serving as vital evidential information. Additionally, the spot mode promptly locates displaced items and facilitates their recovery through the application of structural similarity concepts. These integrated modules reinforce security protocols and ensure rapid responses to unauthorized actions. Furthermore, the system's capabilities extend to the identification of authorized individuals, the recording of suspicious activities, and the simplification of lost item retrieval, all facilitated by the synergy of face training, motion detection, and spot mode.

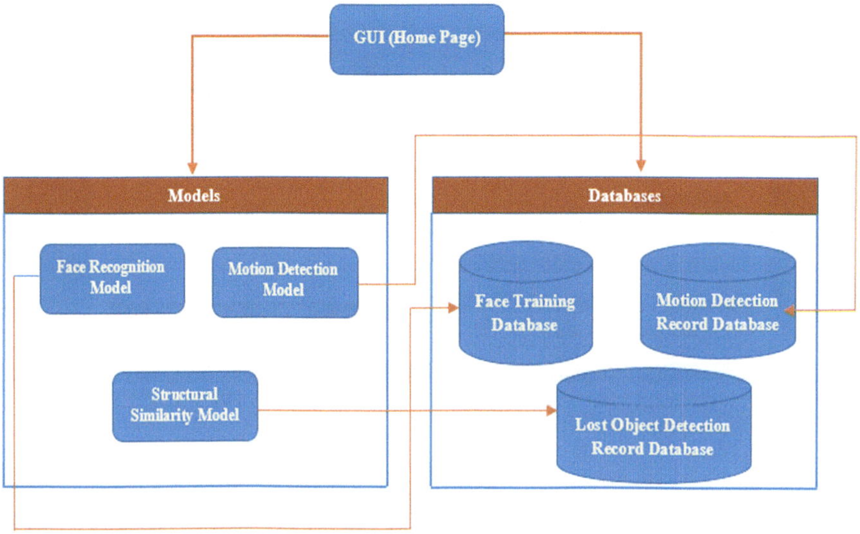

Fig. 1 Architecture diagram of anti-theft system

4 Results and Discussion

A meticulously curated and diverse training dataset is pivotal in developing a Python- and OpenCV-based system for lost object retrieval. This dataset should encompass varied images of the subject under differing settings, lighting conditions, and angles, enhancing the algorithm's accuracy across real-world scenarios. These images illuminate the application's functionality. Collectively, this system offers a comprehensive solution for lost object identification, face recognition, and motion detection. The central user interface, depicted in Fig. 2, serves as the starting point for users searching for lost items, featuring buttons for face training, input fields for IDs and names, and timestamps. Users can link specific IDs and names to activities or data, bolstering system functionality.

By pressing the button, face training starts, building a reference model for identification. Logging timestamps enable the tracking of events.

The GUI uses computer vision or image processing to detect lost objects and aid in their recovery. By combining these capabilities, it can be used for a number of things, including inventory monitoring, attendance management, security systems, and access control. A sample of the faces from the dataset that are used for facial recognition is shown in Fig. 3. These pictures help the algorithm learn to recognize people who might be connected to the disappearance of the object. The database of the facial recognition system contains 50 examples of faces. These samples are essential for identifying authorized people. The system creates a solid reference library for comparison and verification by taking and archiving a sizable number of face samples. In order to assess whether a person is authorized, the system uses

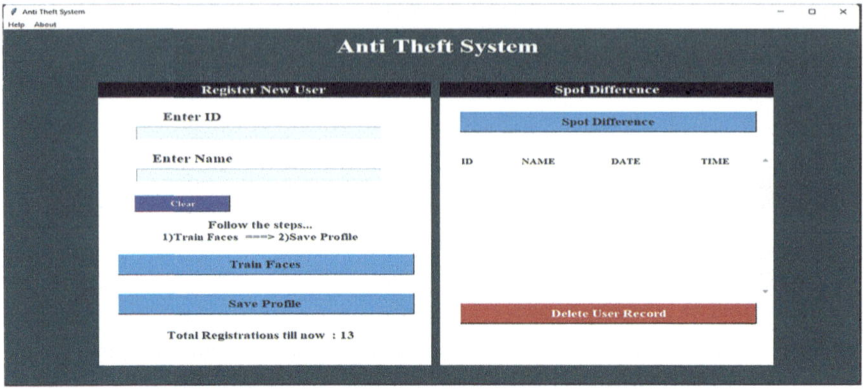

Fig. 2 GUI of the application

Fig. 3 Database of the faces for the training model

facial recognition to compare the captured face to the samples that have been stored. This strategy improves the facial recognition process' accuracy and dependability, making it possible to quickly identify authorized people. The database provides the system with a useful resource for referencing and comparing faces, ensuring secure and trustworthy access control.

The technology effectively detects motion triggered by unauthorized individuals attempting to remove objects, ensuring prompt identification and recording of any suspicious access. This functionality serves to heighten security measures by vigilantly monitoring and documenting such incidents, thereby enhancing the system's capability to identify potential security breaches. The illustration of motion detection is portrayed in Fig. 4, depicting an intruder in the act of attempting to steal an item. The system's motion detection algorithm is adept at spotting such dubious activities, subsequently initiating the necessary actions for intervention.

Fig. 4 Motion detected and recording in action

Figure 5 showcases the application's outcome, highlighting its successful detection of the lost object pilfered by the intruder. This achievement underscores the system's efficacy in locating and pinpointing the object, a crucial aspect in facilitating the recovery process. By synergizing comprehensive training data, facial recognition technology, motion detection algorithms, and object detection methods, the system excels in precise and reliable lost object identification. This amalgamation amplifies its practicality and relevance in real-world scenarios, empowering users to locate their misplaced belongings with confidence.

5 Conclusion

When a face or item is in the camera's range of view, the system detects it and compares it to a database of authorized people. The technology can sound an alarm or notify the authorities if a match is discovered. To prevent any potential exploitation of the obtained data, it is crucial to stress that such systems must be designed while keeping privacy and ethical considerations in mind. In conclusion, the structure similarity-based anti-theft system is a viable method for finding misplaced items using Python and OpenCV. The system compares photographs of missing objects with images in a database using computer vision techniques, and it can precisely identify a match based on the structural similarity index. In addition to providing a reliable method for object detection, the structural similarity technique for image processing makes use of Python and OpenCV. The system is adaptable and can be tailored to meet various contexts and item kinds, making it a flexible option for a variety of situations. Like every system, there may be drawbacks and difficulties to take into account. The requirement for a database of photos to compare against poses a problem because it could take a lot of time and effort to build and maintain. Additionally, the system's accuracy may be affected by elements like illumination

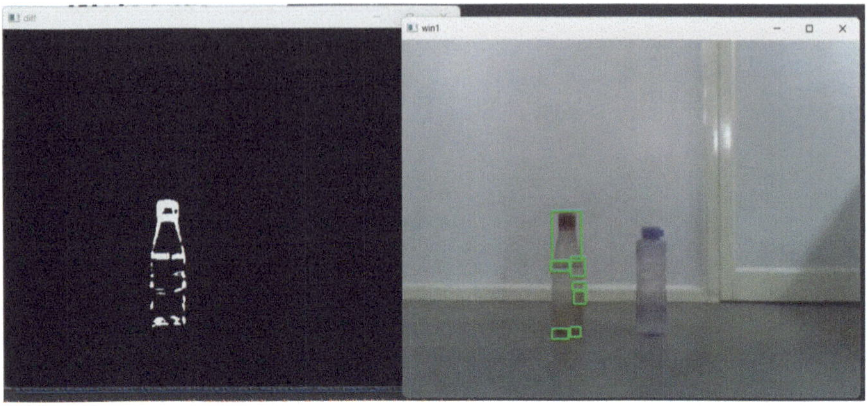

Fig. 5 Identifying the stolen object

and image quality. Overall, the structural similarity-based anti-theft system provides a promising method for finding lost items and has the potential to be a useful security and loss prevention tool. It has the potential to play a significant role in security systems across a range of industries with more research and improvement.

To enhance the current security system and create a comprehensive and reliable anti-theft solution, several key improvements can be implemented. Intelligent object tracking methods could be integrated to monitor and analyze the behaviors of identified objects in real-time. This would significantly improve threat detection and response by enabling the system to identify potentially dangerous situations and raise alerts accordingly. By integrating with facial recognition databases and comparing collected faces to a broader known database, the system's identification accuracy would be greatly enhanced. This feature would be invaluable in identifying potential intruders or individuals with a history of criminal activity. To improve the evaluation of security risks, advanced analytics algorithms could be employed to extract useful information from surveillance data. This could include behavior analysis, crowd density estimation, anomaly detection, and predictive analytics. These insights would allow users to proactively address security issues and prevent incidents before they occur. Integration with Internet of Things (IoT) devices and enhanced object detection algorithms would further strengthen the system's capabilities. The collaboration with the existing security systems and privacy protection measures would ensure seamless integration and compliance with privacy regulations. In conclusion, these enhancements would create a cutting-edge anti-theft system that offers comprehensive surveillance, intelligent threat detection, and real-time response capabilities. The integration of cloud storage, facial recognition, IoT devices, and advanced analytics would make it a robust and reliable solution for safeguarding properties and assets. By combining these features with privacy protection measures and cooperation with existing security systems, the system would offer both efficiency and peace of mind to users, making it a valuable asset in the fight against theft and unauthorized access.

References

1. A. Kumari Sirivarshitha, K. Sravani, K. S. Priya and V. Bhavani, "An approach for Face Detection and Face Recognition using OpenCV and Face Recognition Libraries in Python," 2023 9th International Conference on Advanced Computing and Communication Systems (ICACCS), Coimbatore, India, 2023, pp. 1274–1278, https://doi.org/10.1109/ICACCS57279.2023.10113066.
2. Jiawei Deng, Zhenming Yu, Guangyao Pang, "Colour Variation Minimization Retinex Decomposition and Enhancement with a Multi-Branch Decomposition Network", Chinese Journal of Electronics, vol.32, no.4, pp.908–919, 2023.
3. OpenCV, "2021 Third International Conference on Intelligent Communication Technologies and Virtual Mobile Networks (ICICV)", pp. 1304–1309, 2021, February.
4. Cao Z, Liao T, Song W, Chen Z, Li C (2021) Detecting the shuttlecock for a badminton robot: a YOLO based approach. Expert Syst Appl 164:113833.
5. Zongyu Guo, Runsen Feng, Zhizheng Zhang, Xin Jin, Zhibo Chen, "Learning Cross-Scale Weighted Prediction for Efficient Neural Video Compression", IEEE Transactions on Image Processing, vol.32, pp.3567–3579, 2023.
6. W. Qiao, B. Chen, J. Zhao and S. Guo, "Face Recognition Based on CD-RBM and BMILM", Journal of Physics: Conference Series, vol. 1802, no. 3, pp. 032077, 2021, March.
7. Y. Li, K. Guo, Y. Lu and L. Liu, "Cropping and attention based approach for masked face recognition", Applied Intelligence, pp. 1–14, 2021.
8. N. Damer, F. Boutros, M. Süßmilch, M. Fang, F. Kirchbuchner and A. Kuijper, *Masked Face Recognition: Human vs. Machine*, 2021.
9. M. Loey, G. Manogaran, M. H. N. Taha and N. E. M. Khalifa, "A hybrid deep transfer learning model with machine learning methods for face mask detection in the era of the COVID-19 pandemic", *Measurement*, vol. 167, pp. 108288, 2021.
10. Z. Cheng, H. Sun, M. Takeuchi and J. Katto, "Learned image compression with discretized Gaussian mixture likelihoods and attention modules", Proc. IEEE/CVF Conf. Comput. Vis. Pattern Recognit. (CVPR), pp. 7936–7945, Jun. 2020.
11. Agarwal S, Terrail JO, Jurie F (2018) Recent advances in object detection in the age of deep convolutional neural networks. arXiv preprint arXiv:1809.03193.
12. K. K. Kumar, H. Nat raj and T. P. Jacob, "Motion activated security camera using Raspberry Pi", In 2017 International Conference on Communication and Signal Processing (ICCSP), pp. 1598–1601, 2017, April.
13. Nikolaos Stekas, Dirk van den Heuvel. "Face recognition using Local Binary Patterns Histograms (LBPH) on an FPGA-based System on Chip (SoC)", IEEE International Parallel and Distributed Processing Symposium Workshops, 2018.
14. Haindl Michal and Mikes Stanislav, "Texture segmentation benchmark", 19th International Conference on Pattern Recognition, 2018, ISBN 978-1-4244-2174-9.
15. S L Suma, Sarika Raga. "Real Time Face Recognition of Human Faces by using LBPH and Viola Jones Algorithm." International Journal of Scientific Research in Computer Science and Engineering ,Vol.6, Issue.5, pp. 01–03, Oct. 2018.
16. Cao Z, Liao T, Song W, Chen Z, Li C (2021) Detecting the shuttlecock for a badminton robot: a YOLO based approach. Expert Syst Appl 164:113833.
17. Ahmed, Aftab, Guo, Jiandong, Dharejo, Fayaz, Deeba, Farha, Ahmed, Awais, "LBPH Based Improved Face Recognition At Low Resolution", https://doi.org/10.1109/ICAIBD.2018.8396183
18. Li Cuimei, Qi Zhiliang. "Human face detection algorithm via Haar cascade classifier with three additional classifiers", 13th IEEE International Conference on Electronic Measurement & Instruments, pp. 01–03, 2017.
19. ARedmon J, Divvala S, Girshick R et al 2016 You only look once: Unified, real-time object detection[C] In: Proceedings of the IEEE Conference on Computer Vision and Pattern Recognition 779–788.

20. Sun, Yan, Ren, Zhenyun, Zheng, Wenxi "Computational Intelligence and Neuroscience—Research on Face Recognition Algorithm Based on Image Processing", https://doi.org/10.1155/2022/9224203
21. Tax Niek, Verenich Ilya, La Rosa Marcello and Dumas Marlon, "Predictive Business Process Monitoring with LSTM neural networks", Proceedings of the International Conference on Advanced Information Systems Engineering (CAiSE). Lecture Notes in Computer Science, vol. 10253, pp. 477–492, 2017.
22. Nabian Mohammad Amin and Meidani Hadi, "Deep Learning for Accelerated Reliability Analysis of Infrastructure Networks", Computer-Aided Civil and Infrastructure Engineering, vol. 33, no. 6, pp. 443–458, August 2017.

Analysis and Prediction of CO_2 Emissions, Energy Use, and Economic Growth Trilemma in India

Pratul Singh, Rohan John Santosh, and K. Anitha

1 Introduction

The study aims to predict the growth of GDP in India and whether it can keep up with the GDP of an already developed country, in this instance, China. We use the energy usage data and CO_2 data to identify their correlation with GDP and its growth. We will be extrapolating specific parameters to map the growth rate of developed countries to analyze what price another country has to pay considering the current regime of climate change policies. When we consider China, we can extrapolate parameters like usage of fuels, emission of CO_2 which was prevalent when the country was growing economically, and if the same consumption will be feasible in the next 10–20 years of growth in India. We will use the Augmented Dickey-Fuller (ADF) test to analyze the time series available to us. The ADF test allows us to identify the presence of unit roots in the time series. Some stochastic processes have a unit root, which can be problematic when using time-series models for statistical inference. If the root of the characteristic equation of a linear stochastic process is 1, then the process has a unit root. Although such a process is non-stationary, it is not always trendless. Following this, we will be using the Linear Regression Supervised Machine Learning Model to predict the CO_2 emission, energy use, and GDP growth of both the countries, India and China, and identify a correlation between the factors that govern these characteristics. We also use the XGBoost model which is used to scale datasets to facilitate the prediction of such values for the coming years. The research and prediction give us the ability to

P. Singh (✉) · R. J. Santosh · K. Anitha
Department of Computing Technologies, SRM Institute of Science and Technology, Chennai, Tamil Nadu, India
e-mail: ps7115@srmist.edu.in; rs3931@srmist.edu.in; anithak3@srmist.edu.in

A. Haldorai et al. (eds.), *6th EAI International Conference on Big Data Innovation for Sustainable Cognitive Computing*, EAI/Springer Innovations in Communication and Computing, https://doi.org/10.1007/978-3-031-54696-9_12

correlate energy use and CO_2 emission and identify the effects it will have on the growth of the economy and whether they have aproportional effect on the GDP. Also, we obtain the ability to predict the values for CO_2 emissions, energy use, and economic growth beyond the values available in our dataset.

2 Literature Review

The extensive body of literature exploring the connection between CO_2 emissions, energy consumption, and economic growth is both substantial and contentious. Our research findings align with recent studies conducted by Fávero et al. [1] and Khan et al. [2], findings demonstrating a connection between carbon dioxide emissions and worldwide economic growth. Moreover, there has been a noteworthy observation that the industrial progress in many developing and emerging nations has coincided with the use of polluting fuels, contributing significantly to the rise in global warming and, in particular, the escalation of CO_2e levels. The level of CO_2e has now reached an unprecedented level not witnessed in millennia. According to the International Energy Agency's report, in 2022, global CO_2 emissions from energy climbed by 0.9%, or 321 million metric tons, breaking the previous record of more than 36.8 billion metric tons. The analysis also showed that, excluding China, emissions from emerging market and developing economies in Asia increased by 4.2% or 206 million tons in 2022.

The region's emissions increased by more than half as a result of coal-fired power generation. Over half of the region's emissions increase resulted from coal-fired power generation [3]. Onofrei et al. [4], in their paper, use dynamic ordinary least squares (DOLS), unit root tests, and cointegration techniques to undertake an empirical study on the relationship between economic growth and CO_2 emissions using on new data based on a panel of European Union nations.

Most of the studies have explored the link between economic development and CO_2 emissions within the context of discussions surrounding the pursuit of a "sustainable form of growth" that should align with the objective of reducing CO_2 emissions. Azam et al. [5] concluded that based on their examination of environmental deterioration within specific high CO_2 emission economies, China, Japan, and the USA were found to have positive relationships between CO_2 emissions and economic growth. Yousef-Sahzabi et al. [6] corroborated a pragmatic association between CO_2 emissions and economic growth through their examination of Iran. Bouznit and Maria del [7] affirmed similar findings for Algeria, while Lesakova and Ondrej [8] concluded the same for the Czechia. Magazzino [9] emphasized the role of GDP as the driving force for both energy consumption and CO_2 emissions in Israel. Li [10] and Pao and Tsai [11] found that energy consumption exerts a positive and significant effect on CO_2 emissions in the long-term equilibrium for the BRIC nations.

We found that there is a literature gap and no studies relating to GDP growth, carbon dioxide emissions, and energy use per capita that have been based on India and its challenges to become a developed country. Hence, our chapter focuses to bridge this gap by analyzing these parameters for India and try to predict that will it possible for India to navigate through the climate restrictions that have been put in place and also achieve the goal of becoming a developed country.

3 Dataset

Three input variables were utilized for conducting this research: energy use, CO_2 emissions, and GDP as input indicators for China and India. Time series data for China and India were used in this study. The data on energy use was expressed in kWh. The data on CO_2 emissions was expressed in million metric tons, and GDP was expressed in USD.

4 Methodology

4.1 Augmented Dickey-Fuller Test

The Augmented Dickey-Fuller Test model was utilized to determine whether a time series is stationary or not. One of the most fundamental concepts in time-series analysis is stationarity. Stationary time series data has statistical properties that do not change with time. Some examples of these properties are mean and variance. If mean and variance are constant and they do not change with time, the time series is easier to model and analyze.

Key steps in the ADF test were as follows:

1. Formulate hypotheses:
   - Null hypothesis (*H*0): Presence of a unit root in the time series data. The time series is non-stationary.
   - Alternative hypothesis (*H*1): Lack of a unit root in the time series data. The time series is stationary.
2. Estimate the regression model: The ADF test estimates a regression model of the form:

$$y_t = c + \beta t + \alpha y_{t-1} + \phi \Delta Y_{t-1} + e_t \tag{1}$$

The Augmented Dickey-Fuller test also provides facilities to integrate lagged differences:

- No presence of constant but no trend: $\Delta y_t = \gamma y_{t-1} + \sum^{m}_{s=1} a_s \Delta y_{t-s} + v_t$.
- Presence of constant but no trend: $\Delta y_t = \alpha + \gamma y_{t-1} + \sum^{m}_{s=1} a_s \Delta y_{t-s} + v_t$.

- Presence of constant along with trend: $\Delta y_t = \alpha + \gamma y_{t-1} + \lambda_t + \sum_{s=1}^{m} a_s \Delta y_{t-s} + v_t$.

The DF test statistic is:

$$\mathrm{DF}_{\tau} = \frac{\hat{\gamma}}{\mathrm{SE}(\hat{\gamma})} \tag{2}$$

3. ADF test: A test statistic is calculated based on estimated parameters in the regression model. The critical values are compared to this test statistic from statistical tables to determine whether to reject the null hypothesis.
   - If the critical value is greater than the test statistic, the null hypothesis is rejected, which means that the data is stationary.
   - If the critical value is less than the test statistic, the null hypothesis is not rejected, which means that the data is not stationary.

The choice of lag order is a significant consideration when conducting the ADF test. The lag order can be defined as the numbers of lags to be included in the regression model.

4.2 Linear Regression

Linear regression was the supervised machine learning algorithm used for the research and prediction. Linear regression visualizes a linear correlation between independent variables and dependent variables. The main functionality provided by linear regression is the ability to predict the outcome of future events. The variable that serves as a predictor, which is not influenced by changes in other variables, is known as the independent variable. In our case, CO_2 emissions and energy use are utilized as independent variables for the XGBoost algorithm. However, the dependent variable, GDP, in our case, changes when the independent variable fluctuates. The response or result variable under study or analysis is the dependent variable, and the regression model predicts its value.

Some benefits that encouraged us to select linear regression were:

1. Trouble free implementation
2. Scalability
3. Ease of interpretation
4. Efficient modelling

Linear regression can be calculated using:

$$Y = a + bX \tag{3}$$

Coefficients a and b can be calculated using:

$$a = \frac{\left[(\Sigma y)(\Sigma x^2) - (\Sigma x)(\Sigma xy)\right]}{\left[n(\Sigma x^2) - (\Sigma x)^2\right]} \tag{4}$$

$$b = \frac{\left[n(\Sigma xy) - (\Sigma x)(\Sigma y)\right]}{\left[n(\Sigma x^2) - (\Sigma x^2)\right]} \tag{5}$$

where

x and y are the variables which are used to model the regression line
b denotes the slope of the line
a is the Y-intercept of the line
X are the first dataset values
Y are the second dataset values

4.3 XGBoost Algorithm

Additionally, the XGBoost Algorithm was used to help understand the data. A powerful machine learning technique, XGBoost can aid in improving decisions. An application of gradient-boosting decision trees, XGBoost can be used to improve models for machine learning. Weights are significant in XGBoost. Each independent variable is given a weight before being fed into the decision tree that forecasts outcomes. Decision trees are created in a sequential form. The factors that the tree incorrectly anticipated are given more weight before being placed into the second decision tree. These distinct classifiers and predictors are then combined to create a robust and accurate model.

Motivation to utilize XGBoost:

1. Standardization
2. Utilization of the weighted quantile sketch approach to handle sparse datasets
3. Analogous learning
4. The capacity to manage missing values permits the management of real-world data without the need for extensive pre-processing.

$$\begin{aligned} f(x) &\approx f(a) + f'(a)(x-a) + \frac{1}{2} f''(a)(x-a)^2 \\ \mathcal{L}^{(t)} & \quad \sum_{i=1}^{n} \left[l\left(y_i, \hat{y}^{(t-1)}\right) + g_i f_t(x_i) + \frac{1}{2} h_i f_t^2(x_i) \right] + \Omega(f_t) \end{aligned} \tag{6}$$

Initially, the Augmented Dickey-Fuller test has been utilized to measure the trend-stationarity of the time series. Once the data has been processed, it is available to use for prediction. We opted to use the linear regression model for prediction of the economy. First, we plot the trends in previous years for both CO_2 emissions and energy use, for both countries India and China. Now that this data can be extrapolated and is fed into the XGBoost Algorithm.

XGBoost Algorithm allows us to scale up our datasets and work with real-world values more efficiently. XGBoost provides the final modeling which allows us to correlate energy use and CO_2 emissions to GDP growth in India and China and allow us to draw conclusions.

5 Results

The plot compares CO_2 emissions for the country of India from 1965 to 2021. It has been represented in a rudimentary time series model. The graph shows the drop in CO_2 production owing to restrictions placed due to COVID (Fig. 1).

The plot compares energy usage per capita for the country of India from 1965 to 2021. It reflects a slight dip in energy use, as for CO_2 emissions, due to COVID restrictions (Fig. 2).

This time series plot represents the growth in the economy of the country of India from 1965 to 2021. It reflects fluctuations in the economy of India. We can observe a stark drop in the economy because of the COVID-19 pandemic (Fig. 3).

The plot was obtained after performing the ADF test for stationarity, with $\Phi = 0.5$. It returns a p-value <0.5 which indicates the time series is stationary (Fig. 4).

The plot was obtained after performing the ADF test for stationarity, with $\Phi = 1$. It returns a p-value >0.5 (very high) which indicates the time series is non-stationary (Fig. 5).

Both of these graphs were generated for the country of India.

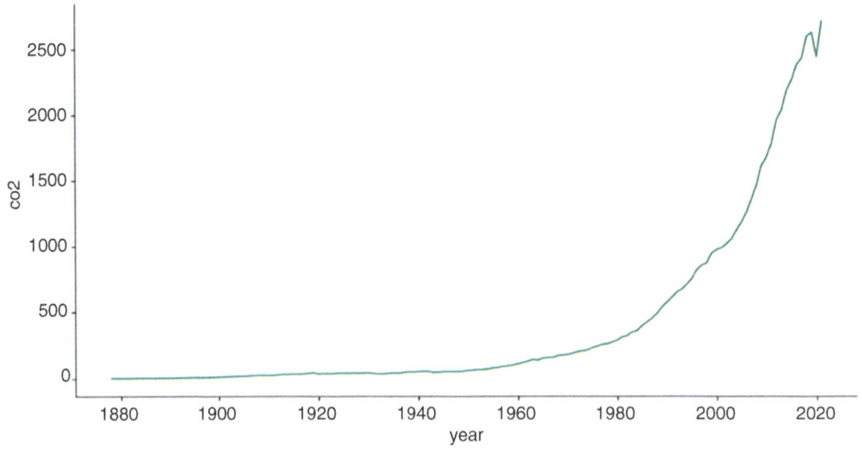

Fig. 1 Drop in CO_2 production owing to restrictions placed due to COVID

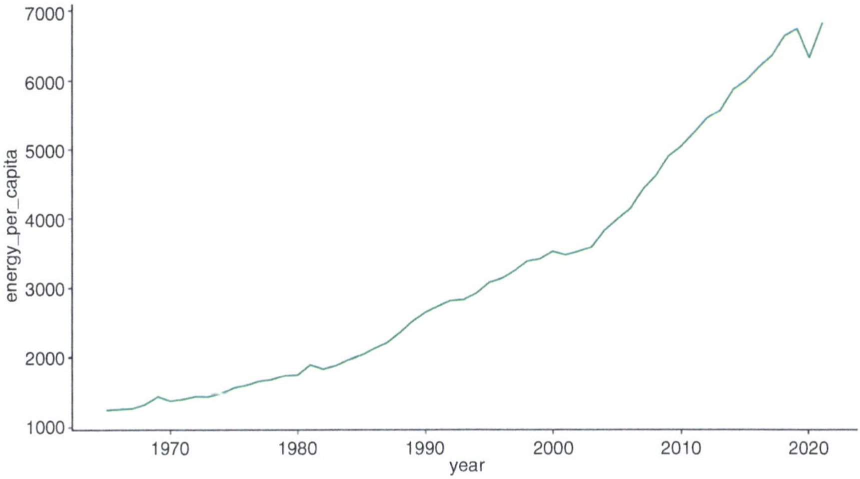

Fig. 2 Slight dip in energy use, as for CO_2 emissions, due to COVID restrictions

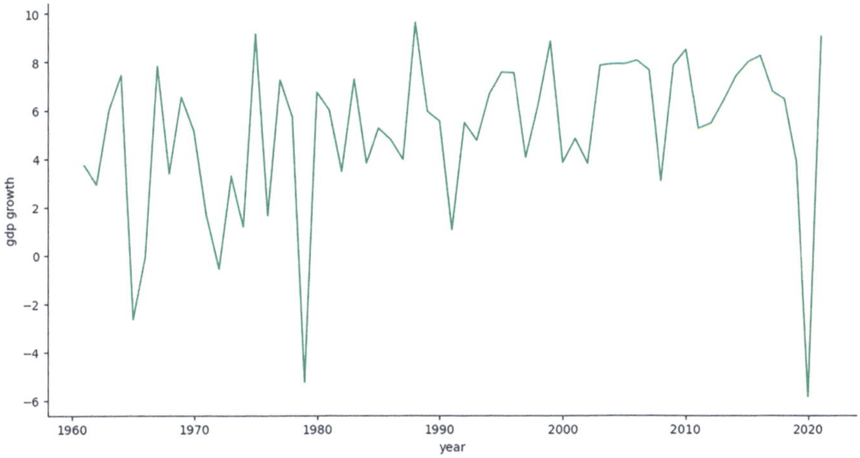

Fig. 3 Reflects fluctuations in the economy of India

The time-series plot represents the CO_2 emission growth in the period from 1965 to 2021 for the country of China. We see a steady rise despite factors like the COVID-19 pandemic (Fig. 6).

The time-series plot represents the growth in energy usage in the period 1965 to 2021 for the country of China. A steady growth can be noticed (Fig. 7).

A time-series plot represents the growth of the Chinese economy in the period from 1965 to 2021. The growth is very noticeable and much more stable than the Indian economy (Fig. 8).

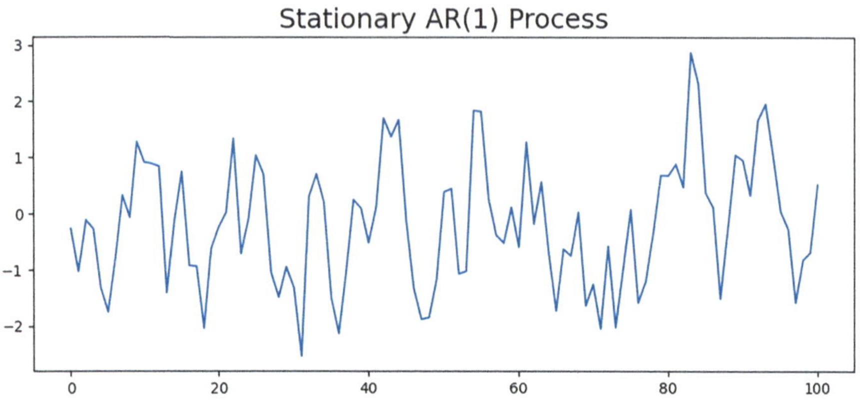

Fig. 4 Time series is stationary

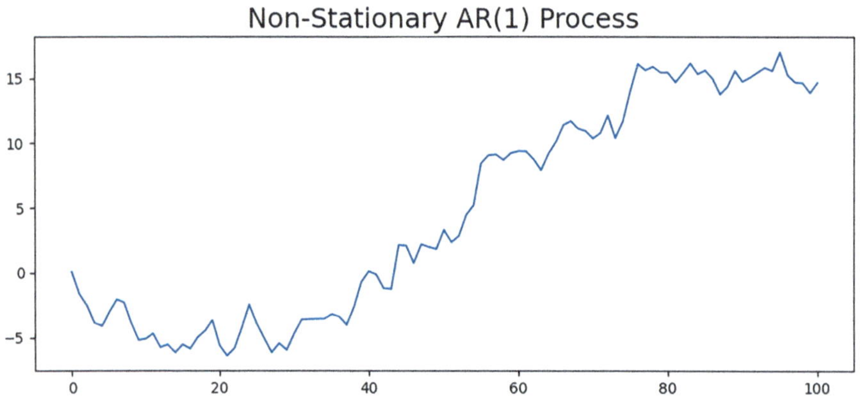

Fig. 5 time series is non-stationary

The plot was obtained after performing the ADF test for stationarity, with $\Phi = 0.5$. It returns a *p*-value <0.5 which indicates the time series is stationary (Fig. 9).

The plot was obtained after performing the ADF test for stationarity, with $\Phi = 1$. It returns a *p*-value >0.5 (very high) which indicates the time series is non-stationary (Fig. 10).

Both of these graphs were generated for the country of China.

Linear regression model implementation to predict energy usage in India till 2020 and beyond is shown in Fig. 11. The graph shows the best-fit line for the data available to the model. One-third of the dataset was used to train the model (Fig. 11).

Linear regression model to predict energy usage in China till 2020 and beyond is shown in Fig. 12. The graph shows the best-fit line for the data available to the model. One-third of the dataset was used to train the model. The plot shows a sharp increase in energy from 2000 (Fig. 12).

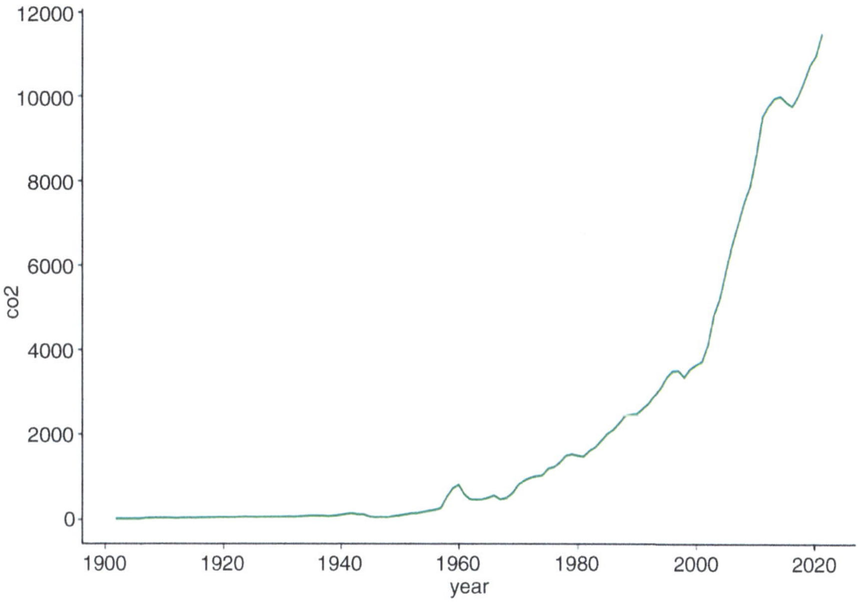

Fig. 6 Steady rise despite factors like the COVID-19

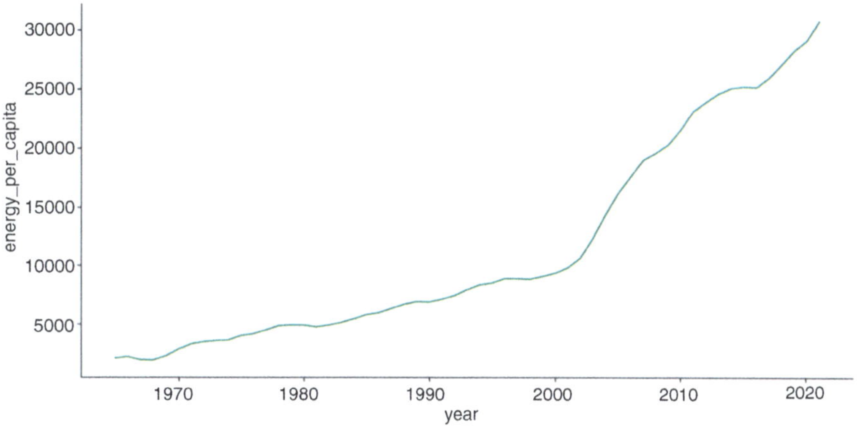

Fig. 7 Steady growth

Linear regression prediction model that maps data for CO_2 emission and extrapolate this data for India from 1965 is shown in Fig. 13. The best-fit line is also represented. A gradual increase is seen in CO_2 emissions (Fig. 13).

Linear regression model to predict CO_2 emissions in China till 2020 and beyond is shown in Fig. 14. The graph shows the best-fit line for the data available to the model. One-third of the dataset was used to train the model. The plot shows a sharp increase in CO_2 Emissions from 2000 (Fig. 14).

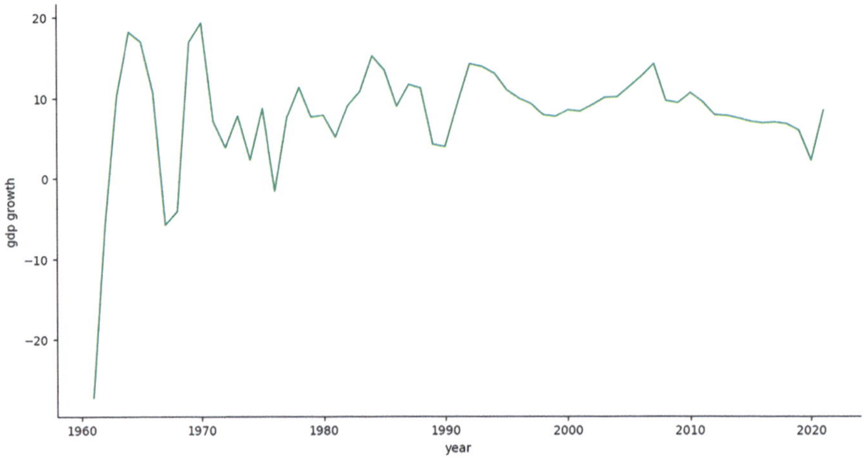

Fig. 8 Chinese economy

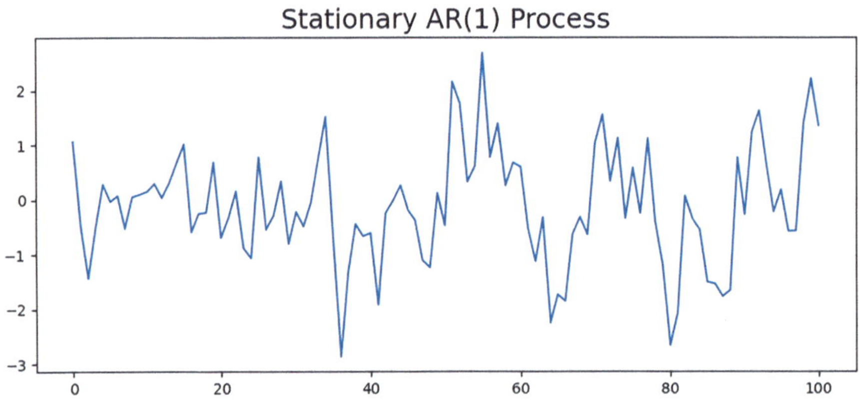

Fig. 9 Performing the ADF test for stationarity

XGBoost algorithm predicting the values of GDP for India is shown in Fig. 15. The values of energy use and CO_2 emissions were plugged in by extrapolation using linear regression. The regression line is also represented for the data (Fig. 15).

XGBoost algorithm predicting the values of GDP for China is shown in Fig. 16. The values of energy use and CO_2 emissions were plugged in by extrapolation using linear regression. The regression line is also represented for the data (Fig. 16).

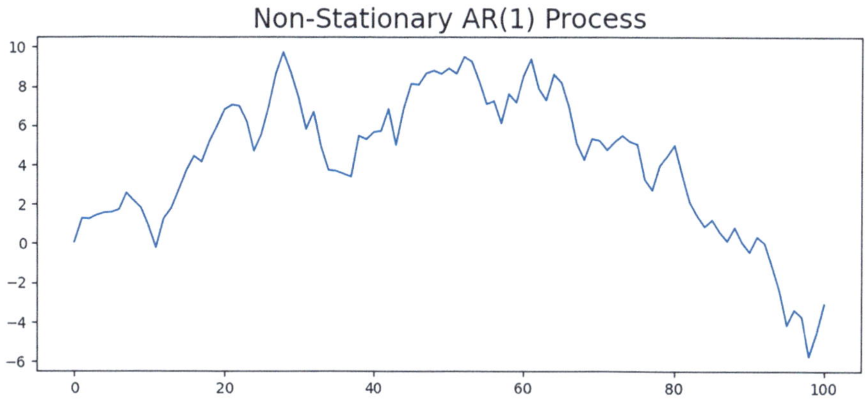

Fig. 10 Series is non-stationary

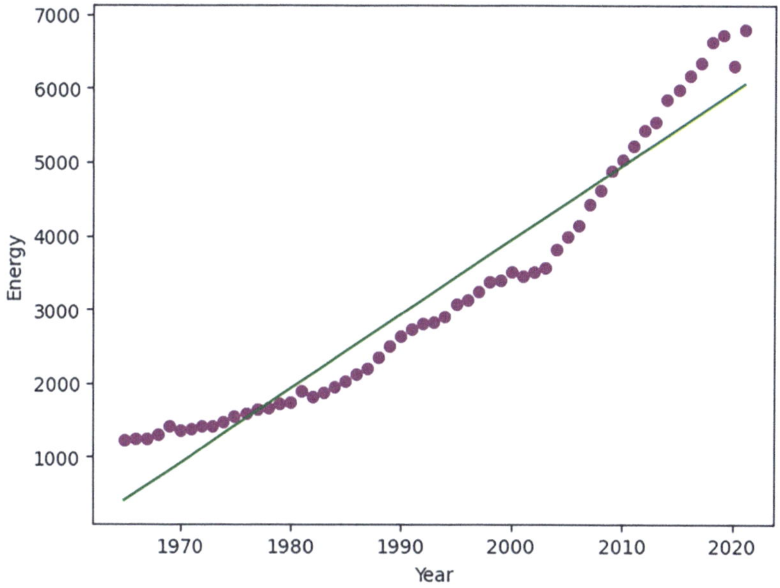

Fig. 11 One-third of the dataset

6 Conclusions

In conclusion to the research, we can say that the GDP of India was vaguely predicted to have remained the same in the next few years or have grown at a proportional rate to energy use and CO_2 emission. Consideration of more factors in prediction model will give us a more accurate result of GDP and its correlation with the factors driving it.

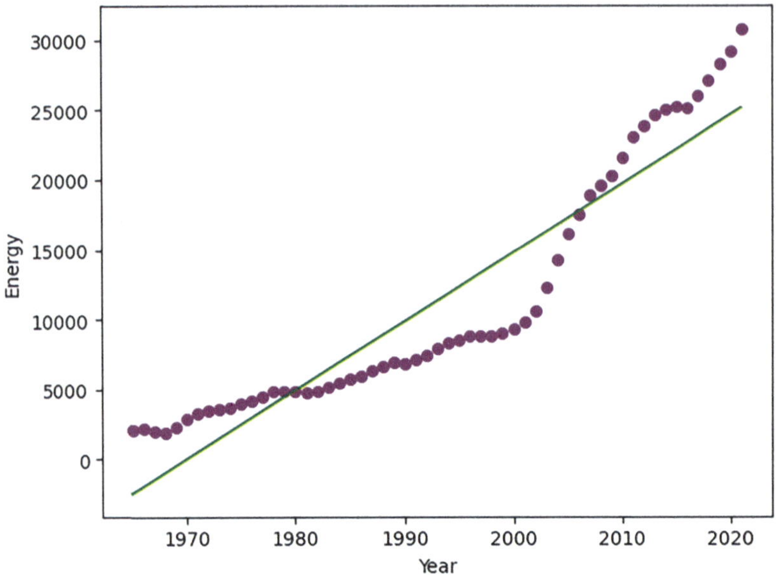

Fig. 12 Sharp increase in energy from 2000

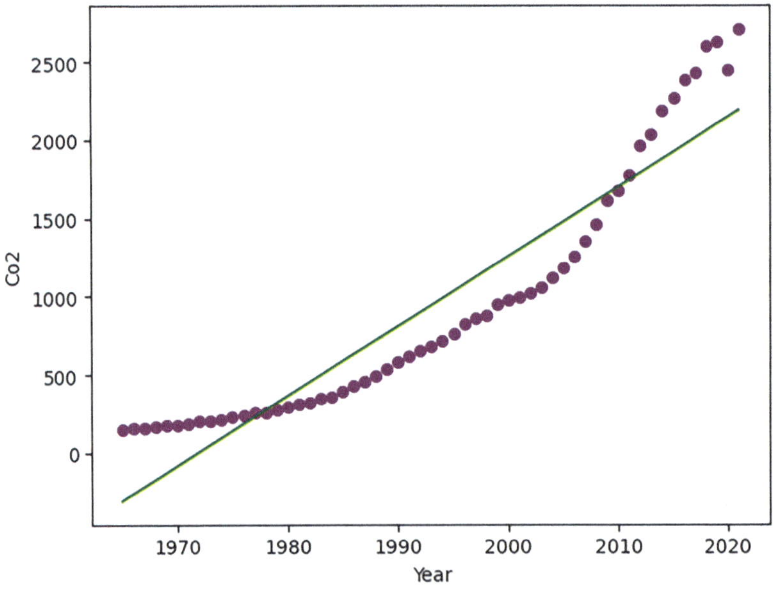

Fig. 13 A gradual increase is seen in CO_2 Emissions

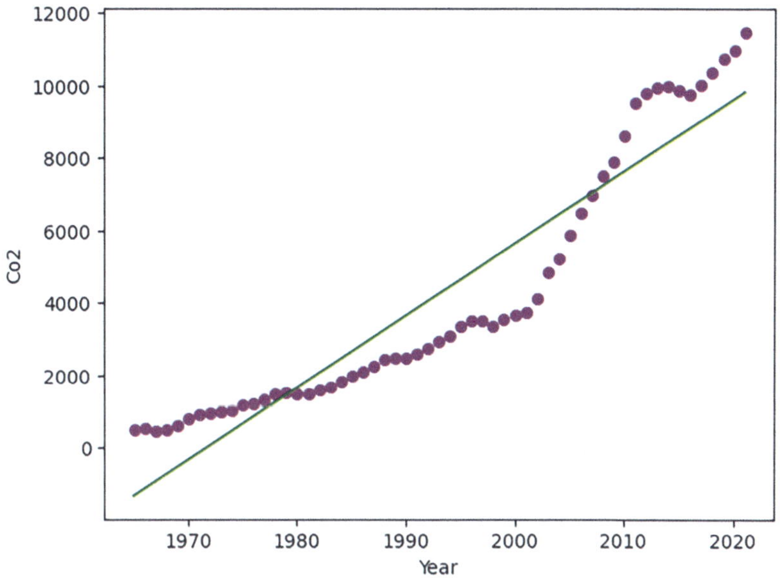

Fig. 14 CO_2 emissions from 2000

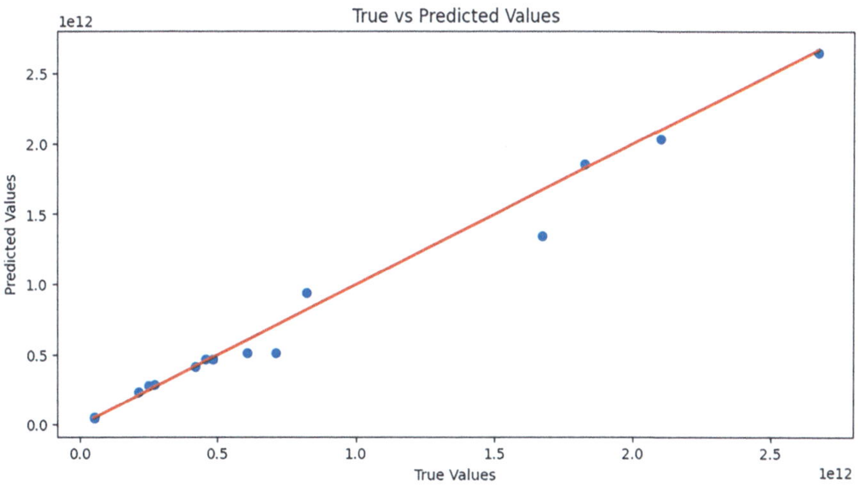

Fig. 15 GDP for India

7 Discussions and Limitations

This chapter aims to fill the void in literature and find the correlation between carbon dioxide emissions, energy use, and economic growth in India. The paper's title mentions the word "trilemma" and rightly so as in the current context it seems

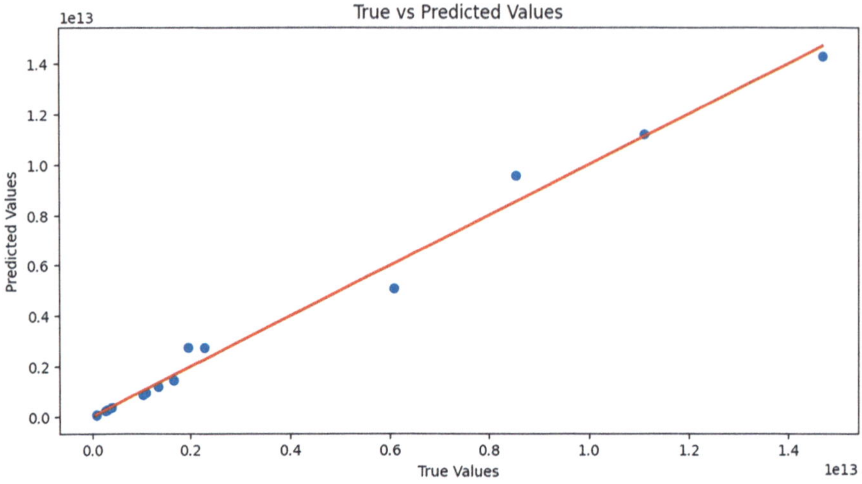

Fig. 16 GDP for China

difficult how any country could manage to honor its commitments made in the United Nations Climate Change Conference of reducing emissions and transitioning to green energy and also maintaining the current levels of growth. What Japan achieved after World War 2 and the economic miracle which has transpired in China showed their reliance on industrialization which means the use of fossil fuels which will increase CO_2 levels and GHG emissions. India's economy is not totally dependent on industries. Services play a significant role in driving Indian economy forward, but as India intends to become an economic superpower the country plans to set up industries in key sectors such as semi-conductors, defense, automobiles, batteries for electric vehicles, etc. This puts India at a tough spot as its aspirations are not aligning with the new global objectives of limiting the use of fossil fuels. In our chapter, we find that if the proportion of energy use and CO_2 predicted using XGBoost remains constant and we only use these two factors in predicting the GDP of both India and China, then our model shows no significant change in the GDP of both the countries by the year 2030, which brings us to the conclusion that only if there is a significant increase in the energy consumption of the country, i.e., investment in industries, and if then all the other factors of the economy see significant growth, then only can India reach its economic goals.

Since our model only takes into consideration CO_2 emissions and energy use as significant parameters, but in reality, GDP growth depends on a number of factors of the economy, it is not showing the true extent of GDP growth; however, it proves that in the current consumption of energy levels and CO_2 emissions, India would maintain its GDP numbers and we find using regression that as energy use increases then GDP also increases. However, if India were to commit to its commitments made in the COP26 in 2021 which stated that there will be a reduction in the carbon intensity of its economy, which compared to 2005 levels, will drop by at least 45%

[12], then it would significantly impact its economic growth as then the country would rely heavily on services which is vulnerable to external shocks. It is evident in the model that higher GDP growth will often lead to more CO_2 emissions. Similar result was found in Ahmed et al. [13]. The authors stated that as income levels increased in India then CO_2 emissions increased; however, USA had a reverse effect, i.e., the CO_2 emissions decreased with increase in economy. Our model also predicts an increase in the CO_2 emissions of India and China over the next decade which will make it difficult for the countries to honor their international commitments.

To get a more precise result, more parameters can be studied using linear regression and then these parameters can be put in XGBoost to get more accurate predictions for the GDP.

References

1. Fávero, L. P., De Freitas Souza, R., Belfiore, P., Roberto Luppe, M., and Severo, M. (2022). Global Relationship between Economic Growth and CO_2Emissions across Time: A Multilevel Approach. Int. J. Glob. Warming 26 (1), 38. https://doi.org/10.1504/IJGW.2022.120067
2. Khan, M. B., Saleem, H., Shabbir, M. S., and Huobao, X. (2022). The Effects of Globalization, Energy Consumption and Economic Growth on Carbon Dioxide Emissions in South Asian Countries. Energy. Environ. 33 (1), 107–134. https://doi.org/10.1177/0958305x20986896
3. IEA (2023), CO_2 *Emissions in 2022*, IEA, Paris https://www.iea.org/reports/CO_2-emissions-in-2022, License: CC BY 4.0
4. Onofrei M, Vatamanu AF and Cigu E (2022) The Relationship Between Economic Growth and CO_2 Emissions in EU Countries: A Cointegration Analysis. Front. Environ. Sci. 10:934885. https://doi.org/10.3389/fenvs.2022.934885
5. Azam, M., Khan, A. Q., Abdullah, H. B., and Qureshi, M. E. (2016). The Impact of CO_2Emissions on Economic Growth: Evidence from Selected Higher CO_2 Emissions Economies. Environ. Sci. Pollut. Res. 23 (7), 6376–6389. https://doi.org/10.1007/s11356-015-5817-4.
6. Yousefi-Sahzabi, A., Sasaki, K., Yousefi, H., and Sugai, Y. (2011). CO_2 Emission and Economic Growth of Iran. Mitig. Adapt Strateg. Glob. Change 16 (1), 63–82. https://doi.org/10.1007/s11027-010-9252-z
7. Bouznit, M., and Pablo-Romero, M. d. P. (2016). CO_2 Emission and Economic Growth in Algeria. Energy Policy 96, 93–104. https://doi.org/10.1016/j.enpol.2016.05.036
8. Lešáková, P., and Dobeš, O. (2018). "Economic Growth and CO_2 Emissions in the Czech Republic," in Vision 2020: Sustainable Economic Development and Application of Innovation Management from Regional Expansion to Global Growth. Seville, Spain
9. Magazzino, C. (2015). Economic Growth, CO_2 Emissions and Energy Use in Israel. Int. J. Sustain. Dev. World Ecol. 22 (1), 89–97. https://doi.org/10.5539/jsd.v8n9p89
10. Li, F., Chang, T., Wang, M-C., and Zhou, J. (2022). The Relationship between Health Expenditure, CO_2 Emissions, and Economic Growth in the BRICS Countries—Based on the Fourier ARDL Model. Environ. Sci. Pollut. Res. 29, 1–20. https://doi.org/10.1007/s11356-021-17900-w
11. Pao, H.-T., and Tsai, C.-M. (2010). CO_2 Emissions, Energy Consumption and Economic Growth in BRIC Countries. Energy policy 38 (12), 7850–7860. https://doi.org/10.1016/j.enpol.2010.08.045

12. Assessing India's Ambitious Climate Commitmentsby Kaushik Deband PranatiCheshta Kohli. Download:—https://www.energypolicy.columbia.edu/publications/assessing-india-s-ambitious-climate-commitments/
13. Ahmed, M.; Huan, W.; Ali, N.; Shafi, A.; Ehsan, M.; Abdelrahman, K.; Khan, A.A.; Abbasi, S.S.; Fnais, M.S. The Effect of Energy Consumption, Income, and Population Growth on CO_2 Emissions: Evidence from NARDL and Machine Learning Models. Sustainability 2023, 15, 11956. https://doi.org/10.3390/su151511956

Cardless ATM Transaction System

P. Sugantha Priyadharshini, A. D. Janani, N. Jayakiruba, and R. Hemalatha

1 Introduction

In today's world, personal data security is a major issue for the average person in the modern world. Personal information leaks result in instantaneous financial ruin; therefore, this enables a secure transaction in ATM systems. Fraud of all kinds is rising rapidly in India. In India, the number of bank account and ATM users is rising quickly, yet fraud in this system is also rising.

So, our aim is to develop a system to increase the ATM system's security. This technology is inexpensive and incredibly safe. Giving a live photo throughout the registration process requires just one more step. People can use a face that is distinctive, simple, secure, and accurate for identification for security purposes. We can also provide the consumer with a benefit in this system by granting access to the account. Several people lose their money in India as a result of ATM security issues. So, it is decided to create software that enhances ATM security [11].

In this chapter, it is suggested a very safe and user-friendly system. Transactions at the ATM system are simple with this method. Face recognition can be used to make the ATM system more secure. A shuffle keypad can also be used as a defense against shoulder surfing assaults. The face can serve as a unique identification for authentication purposes. This method offers a safe transaction and entirely prevents fraud in ATM systems. In a normal ATM transaction, the first card is inserted then the PIN is asked to access the bank account [14]. In the system developed, the first

P. Sugantha Priyadharshini (✉) · A. D. Janani · N. Jayakiruba · R. Hemalatha
Department of CSE, Sri Ramakrishna Engineering College, Coimbatore, Tamil Nadu, India
e-mail: sugantha.ravee@srec.ac.in; janani.2001057@srec.ac.in;
jayakiruba.2001058@srec.ac.in; hemalatha.2001053@srec.ac.in

A. Haldorai et al. (eds.), *6th EAI International Conference on Big Data Innovation for Sustainable Cognitive Computing*, EAI/Springer Innovations in Communication and Computing, https://doi.org/10.1007/978-3-031-54696-9_13

face is recognized. After successful recognition of the user's face, PIN is asked to be entered. After entering the PIN, bank account can be accessed.

If the face recognized doesn't match with the user, then the access request is denied. It appears that this issue is more regrettable as anybody can get all data put away when they enter the right secret word towards getting to an ATM card at ATM. Other than that, it is firmly stressed that the security issues need innovation upgrades and better security strategy as a countermeasure. The working of ATM has the end goal that when a client shows his/her face in front of the camera, it consequently checks to approve [10]. Face recognition can be characterized as a quantifiable physiological and social trademark that can be caught and, in this manner, contrasted and another case at the hour of confirmation. Using face recognition enhances the security of ATM transactions.

2 Objective of the Work

The objective of this research is to develop a cardless ATM transaction system that makes use of facial recognition software to offer a quick and secure way to access ATM services. The technology attempts to lower the danger of theft or loss by doing away with the requirement for a physical bank card and by providing a different way to conduct transactions. The system's significant goal is to provide security and accessibility for users, particularly people with impairments or mobility challenges who might find it challenging to use a physical bank card [9].

A facial recognition module, a realtime database, and an ATM transaction system make up the system's three primary parts. The realtime database's stored facial traits are compared to the user's unique features, which are extracted from their facial image by the facial recognition module. If there is a match, the user is confirmed and permitted to use the ATM transaction system to carry out transactions.

To validate the transaction and offer details on the account balance or other pertinent information, the system sends a notification to the user's mobile device or email address. The system's advantages, such as improved security, usability, and accessibility, make it a desirable choice for customers looking for a hassle-free and secure way to access ATM services. In conclusion, the goal of the work is to create a state-of-the-art cardless ATM transaction system that provides a safe and practical way to conduct transactions utilizing facial recognition technology.

3 Existing System

The ATMs of the modern era use PIN-based security [7]. The PIN is taken into consideration as the information that is encrypted on the customer side and the information is decoded on the server side as we approach the completion of the exchange. We can complete the swap at the time the correlation is satisfied. The card and PIN-based architecture used by the present ATM framework confirms trades. Following that, it grants bank customers access to several services, including cash withdrawal and storage, record-to-account transfers, balance inquiry, top-up purchases, and payment of service fees. The ATM system compares the entered PIN to the put-away authorization PIN for each ATM user.

If there is a match, the system verifies the client and grants access to all services provided by the ATM. If there is an error, the client verification process fails, and the customer is given two more attempts to enter the proper PIN. If an incorrect PIN is entered three times, the ATM will block the card and hold it. Passing the appropriate PIN is sufficient to authenticate a client to the bank's withdrawal framework and grant access to it [8]. The existing system keeps ATM cards after around three wrong PIN entries; after that, it automatically rejects any additional efforts to get access as being unauthorized.

4 Methodology Used in Proposed System

The goal of the proposed study is to create an image processing-based facial recognition system that would allow users to access ATMs without requiring ATM cards. We will therefore be able to conduct transactions quickly and securely by using this method instead of needing to insert our credit cards. It could be challenging to carry the card with us all times, and using the machine becomes too challenging if we forget the card. To avoid this, a technology will be developed to recognize the user's face and enable them to access their bank account. To access the bank account when using a regular ATM, the card must be inserted into the ATM before entering the PIN. After successful facial recognition, the system prompts the user to enter a PIN in order to access their bank account, hence enhancing security by guaranteeing that only the account owner may access the account. ATM cards are no longer necessary thanks to this method, which also makes the account accessible even in the event that the card is misplaced. Users benefit from convenience as facial recognition technology shortens the time needed to access and conduct account transactions. In order for the user to access their bank account, the system needs to successfully recognize their face and enter a PIN. Since only the account owner's face is entered into the database, the system maintains security by limiting access to the account solely to them. This approach offers a different way to access the account and eliminates the need for ATM cards. The time needed to access the account and complete transactions is shortened using face recognition. By enabling face recognition for account access, the system improves accessibility for users.

To develop a cardless ATM transaction system that detects and identifies the user through facial recognition methodology and fetch the account details of the specific user to perform various ATM transactions like cash deposit, cash withdrawal and balance inquiry, the following system is proposed.

Define the problem: Start by defining the problem statement and the scope of the digital transaction. What is the purpose of the system, and what kind of information do you want to gather? Who will be using the system, and what are their requirements?

Database setup: Identify the data sources and collect the images of the user that will be used to recognize and identify the user. Set up a database with the account details like account number, account type, and the total available balance of the user with the specific user ID in the Firebase.

Facial recognition: Link the database with the facial recognition system and once the system recognizes the user, the respective account details of the user will be displayed and further process will be proceeded based on the account details.

Performing ATM transactions: Facial recognition is integrated with the ATM transaction system and realtime database to provide a secure and convenient method of verifying the identity of users and authorizing transactions. This eliminates the need for ATM cards, reduces the risk of fraud, and updates transaction data continuously.

The hardware requirement includes webcam and software requirements like Visual Studio Installer, Firebase, PyCharm, and some of the libraries to be used for our work (Fig. 1).

5 Module Description

5.1 Face Recognition

In this module, facial recognition using a realtime database for ATM transactions is implemented using OpenCV in PyCharm community editor which is an image processing algorithm to verify the identity of individuals making transactions at an ATM machine by comparing their faces to a database of known faces which are uploaded in the python project in the PyCharm that is updated continuously in realtime [10]. The system typically consists of a camera that captures images of faces, a facial recognition module that processes the images, and a database that stores the facial features of authorized individuals. When an individual approaches the system, the facial recognition module captures an image of their face and compares it to the features of known faces in the database. If there is a match, the system can allow the individual to proceed with their transaction. If there is no match, the transaction may be declined or further verification may be required. The use of facial recognition technology for ATM transactions despite using atm cards can enhance security and prevent fraud by ensuring that only authorized individuals can access the account and make transactions. It can also make the transaction process faster and more convenient for users, as they do not need to insert a card every time to proceed with the transaction process. However, it is also important to consider the privacy implications of using facial recognition technology for ATM transactions

Fig. 1 System architecture

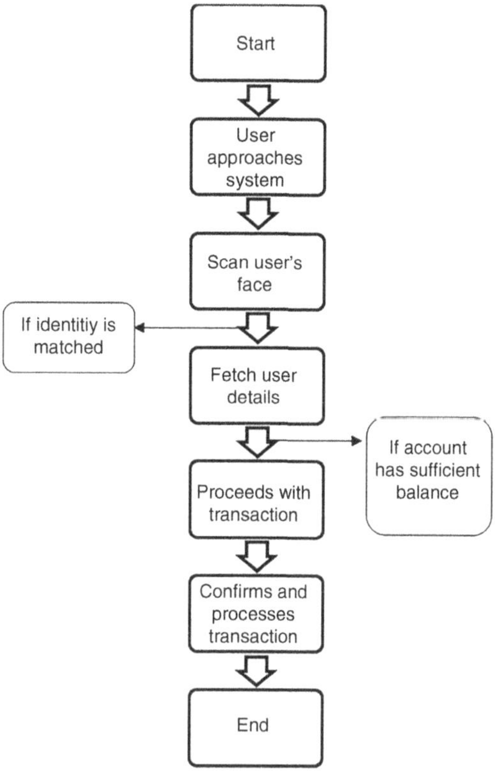

and to ensure that appropriate safeguards are in place to protect individuals' rights and prevent misuse [12].

5.2 *Database Setup*

In this module, setting up a database for facial recognition for performing ATM transactions in Firebase is carried out which involves creating a cloud-based database to store the facial features of authorized individuals and integrating it with the ATM facial recognition system. Firebase is a cloud-based platform that provides realtime database services, making it an ideal platform for this application. To set up the database, the first step is to create a Firebase project and enable the realtime database service. Next, the facial recognition module on the ATM must be configured to capture facial images and extract unique facial features. These features are then compared to the features of known faces in the Firebase database. Authorized users' facial features and the specific person's account details should be stored in the Firebase database as a collection of key-value pairs, with the user's name or ID as the key and the facial feature data as the value. To perform a facial recognition transaction, the facial recognition module on the ATM captures an image of the user's face, extracts the facial features, and compares them to the facial features stored with the respective ID in the Firebase database. If there is a match, the transaction proceeds [13]. If there is no match, the transaction may be declined or further verification may be required. Firebase provides realtime synchronization of data between the database and the ATM transactions, ensuring that the facial feature data is always up to date. This makes it an effective and convenient platform for facial recognition applications, including ATM transactions.

5.3 Performing ATM Transactions

In this module, ATM transactions are performed from a realtime database using facial recognition which integrates the facial recognition module with the ATM transaction system and the realtime database. Here, a UI/UX module is developed to show the ATM transaction processes such as cash withdrawal, balance inquiry, and deposit money. This allows for a secure and convenient method of verifying the identity of the user and authorizing transactions in realtime [14]. The facial recognition module captures an image of the user's face and compares it to the facial features stored in the realtime database. The database contains the facial features of authorized users, which are updated continuously in realtime. The facial recognition module uses advanced algorithms to match the user's face with the stored facial features and verify their identity. Once the user's identity is verified, the ATM transaction system allows them to proceed with the transaction. The user can access their account, withdraw cash, deposit money, or perform any other transaction authorized for their account. All the transaction data are updated continuously to the database and the process proceeded without any inconvenience.

The use of facial recognition for ATM transactions provides a secure and convenient method of identity verification. It eliminates the need for the usage of ATM cards for ATM transactions as they can be compromised or forgotten. Facial recognition also reduces the risk of fraud and identity theft, as it is difficult to replicate someone's facial features.

6 Implementation and Results

Install OpenCV and Firebase Admin in your PyCharm environment. Use OpenCV to access the webcam and capture images of customers. This system uses the Haar Cascade classifier for face detection and extracts the facial features using a pre-trained model like Open Face or dlib. Store the facial features of each customer in a Firebase database. You can also store other relevant information such as the customer's name, account number, and balance. Train machine learning model to recognize faces based on the extracted facial features. Our work uses classifiers like

SVM or KNN for this task. Label the captured images with the respective customers' information and store them in the database. When a customer wants to perform transactions, use OpenCV to capture their image and extract their facial features. Use the machine learning model to recognize the customer based on their facial features and retrieve their information from the database. Authenticate the customer's ID using their password or any other authentication method you have in place. Perform the requested transaction, such as a withdrawal or balance inquiry, by accessing the customer's information in the database. Log the transaction details, including the customer's ID, transaction type, and amount, in the database for future reference. Figure 2 shows the facial recognition.

Firebase is used in order to store the realtime database to the admin login. Realtime database is being created in the Firebase. The images and data fed are being stored in the Firebase's storage. Firebase storage contains all the data required for the account and images of the client which is fed to Firebase admin as depicted in Fig. 3.

A real-time database is a database that is obtained in a realtime source and then fed to the database matched with the recognizing face and checked whether it matches with the one in a database or not. The Firebase realtime database is the cloud-hosted database. The Firebase realtime database is a cloud-hosted NoSQL database that lets you store and sync data between your users in realtime. Figure 4 shows the realtime database.

Figure 5 depicts how the face is recognized once the webcam is on.

Once the face is recognized, the system starts its login procedure for further processing which is depicted in Fig. 6.

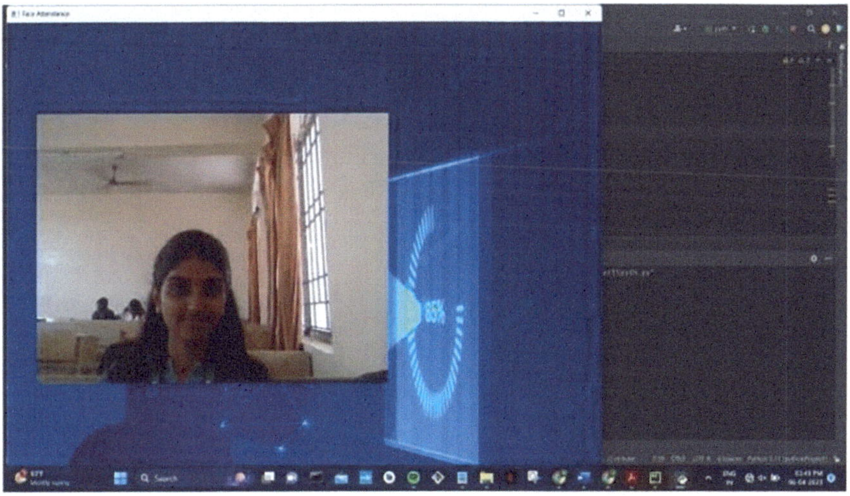

Fig. 2 Face recognition

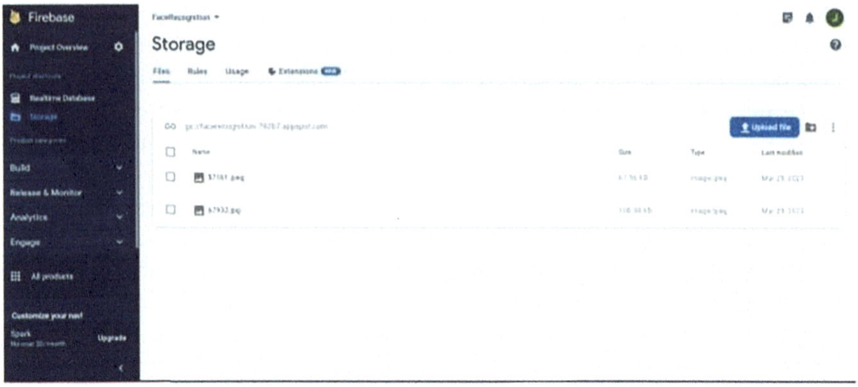

Fig. 3 Firebase

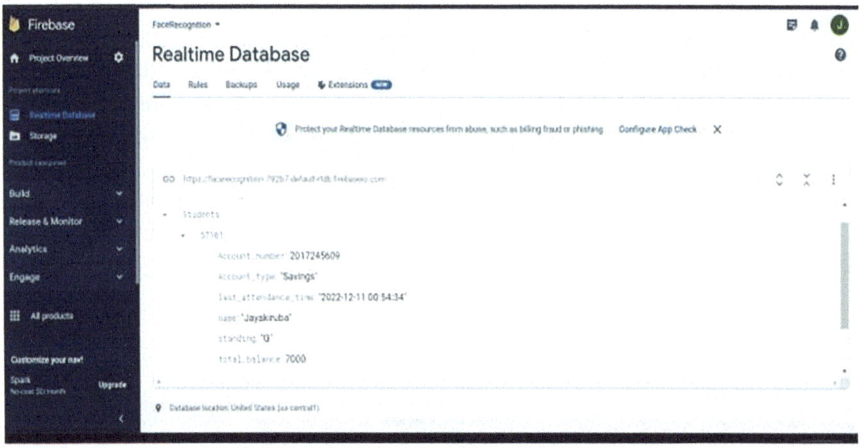

Fig. 4 Realtime database

Once the user is logged in, then the systems move to the transaction procedure where the user can withdraw and deposit in their account without the ATM card. Figure 7 shows the transaction details.

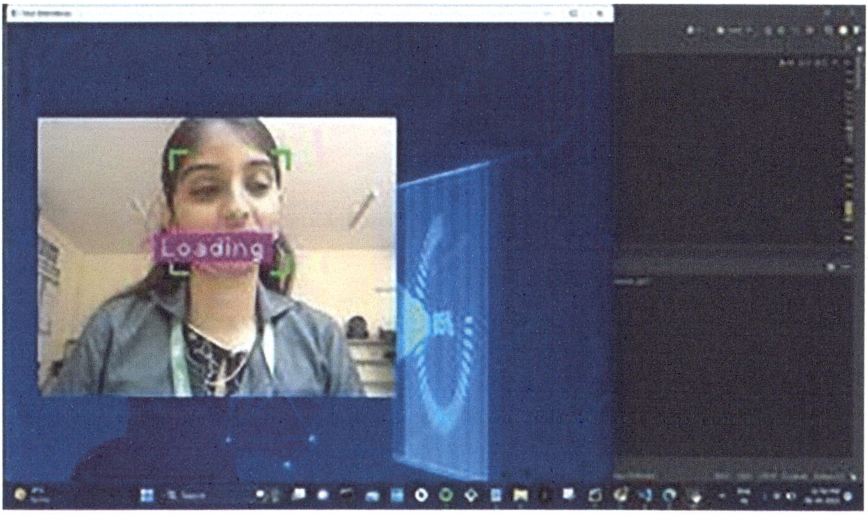

Fig. 5 Recognizing face

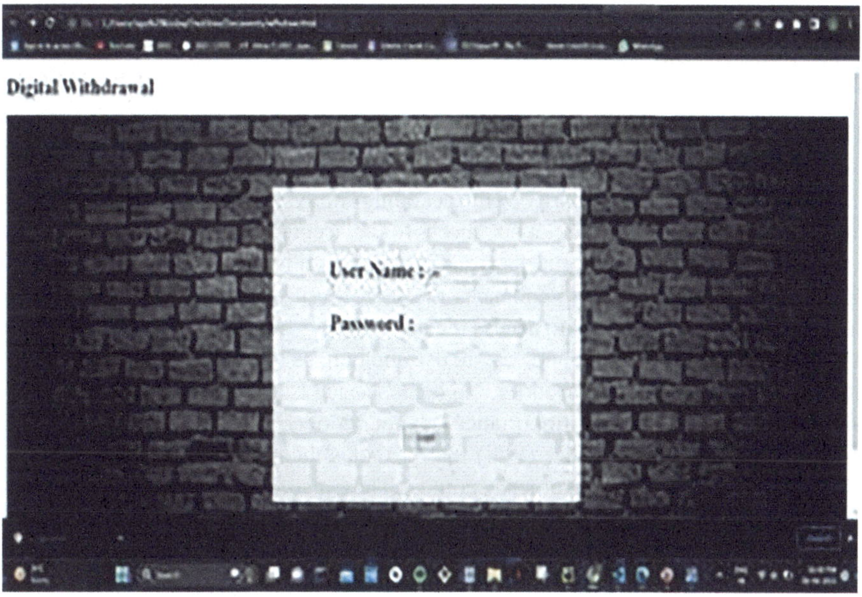

Fig. 6 Login procedure

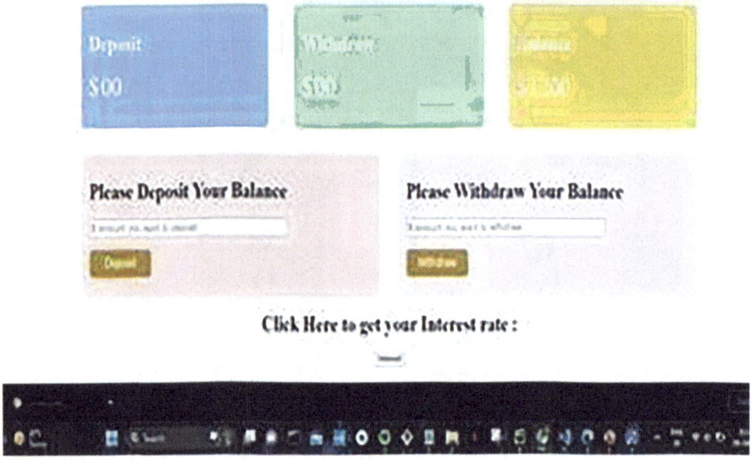

Fig. 7 Transaction details

7 Conclusion and Future Enhancements

7.1 Conclusion

Thus, the system developed will allow the client to access the account by face recognition successfully. This system of recognizing the face and letting the user access the account enhances the security of the account. The system works on face recognition; therefore, fraudsters cannot get access to the account by stealing the cards and knowing the password. The success rate of the system is good; therefore, it can be applied to the real world and enhances the security of the account. Despite using ATM cards and security PIN to access the account, this method of using face recognition is reliable, affordable, and easy to use and help us to access the account easily and quickly with enhanced security. No bank service is as convenient, effective, or efficient as ATMs, so their importance cannot be overstated. As each user has a personal identification number (PIN) that provides them access to their accounts, ATMs offer security to consumer accounts. They also offer banks a quicker and more accurate way to manage data. Since the PIN can be stolen or coerced from the owner by shadowy figures, using it at an ATM does not guarantee the security of the customer's account. When cardless ATM transactions are implemented in such circumstances, the issues associated with ATM card usage can be eliminated, ensuring secure and quick transactions. Cardless ATM transaction not only improves efficiency but also reduces the crowding in the customer care section where complaints about ATM card damage. Card-based ATMs suffer numerous security risks, including card cloning, card damage, card expiration, cast skimming, cost of issuance and maintenance, third parties being able to access client accounts, and waiting periods before issuing expired or damaged cards. The system proposed thus eliminates the above-mentioned issues and ensures secure ATM transaction in a quick and efficient manner.

7.2 Future Enhancement

The future enhancement of the work is to enhance the security by adding the one-time password verification as a step next to face recognition. One-time password verification will enhance the security of the access as the one-time password is sent only to the registered mobile number of the client. Then as an add-on, iris recognition can also be added, which will enhance the security still more. The iris recognition is used to match the iris of the client fed to the database with the recognized one which is recognized at the time of transactions. Iris recognition is also used to find whether a person is forced to stand to get the face recognized by iris dilation. The one-time password verification and iris recognition will enhance security and efficiency. This method will offer the best result compared to the existing systems for ATM transactions. By using the system, we can eliminate fraudulent activities.

References

1. Sahil Bajaj, Sumit Dawda, Pradnya Jadhav, Rasika Shirude "Card less ATM Using Deep Learning and Facial Recognition Features," 2022 Journal Information Technology and Software Engineering, Belgium, 2022, Vol.12
2. Ashwini C, Shashank P, Shreya Mahesh Nayak, Siri Yadav S, Sumukh M, 2020, "Cardless Multi-Banking ATM System Services using Biometrics and Face Recognition," International Journal of Engineering Research & Technology (IJERT) NCCDS-2020, Vol.8
3. Vaishnavi M. Shinde, Dhanashri B. Shinde, Pravin B. Shinde, Aditya V. Wandhekar, Pallavi S. Kohakade, 2023, "Virtual ATM through Fingerprint and Face recognition using Deep Learning", International Journal of Innovative Research in Technology (IJIRT), 2023, Vol 9.
4. N. N. Nandhini, and G. Dinesh, "Cardless ATM Transaction Using Mobile," 2017 International Conference on Innovations in Information, Embedded and Communication Systems (ICIIECS), Coimbatore, 2017, pp. 1–5
5. S. K. Mandal, and S. K. Singh, "Cardless ATM Transactions with Mobile Devices," 2019 11th International Conference on Communication Systems & Networks (COMSNETS), Bengaluru, India, 2019, pp. 646–647.
6. M. Al-Ghazali, and M. Al-Abachi, "Design of a Cardless ATM Transaction System Based on Fingerprint Recognition," 2019 International Conference on Computer and Information Sciences (ICCIS), Najaf, Iraq, 2019, pp. 1–6.
7. S. T. V. R. Manikanta, and S. R. Krishnareddy, "Cardless ATM Transactions Using QR Code and Mobile Authentication," 2018 Second International Conference on Intelligent Computing and Control Systems (ICICCS), Madurai, India, 2018, pp. 932–936.
8. Mohanty, and B. N. Panda, "Cardless ATM Transactions Using Mobile App with Biometric Authentication," 2019 IEEE 3rd International Conference on Trends in Electronics and Informatics (ICOEI), Tirunelveli, India, 2019, pp. 1202–1206. R. K. Patel, and P. B. Patel, "Cardless ATM Transactions Using Mobile Authentication System," 2019 3rd International Conference on Electronics, Communication and Aerospace Technology (ICECA), Coimbatore, India, 2019, pp. 1589–1593.

9. S. S. Sakhare, and R. S. Deokar, "Cardless ATM Transactions Using Mobile Authentication with OTP," 2019 International Conference on Computer Communication and Informatics (ICCCI), Coimbatore, India, 2019, pp. 1–5.
10. N. Singh, and D. D. Kumar, "Cardless ATM Transactions Using QR Code," 2019 4th International Conference on Information Systems and Computer Networks (ISCON), Varanasi, India, 2019, pp. 275–278.
11. S. K. Singh, and S. K. Mandal, "Cardless ATM Transactions with Mobile Devices Using Bluetooth and QR Code," 2019 IEEE International Conference on Advanced Networks and Telecommunications Systems (ANTS), Goa, India, 2019, pp. 1–6.
12. D. V. Thakare, "Cardless ATM Transaction System," 2021 3rd International Conference on Advances in Electronics, Computers and Communications (ICAECC), 2021, pp. 1–5.
13. K. Yadav, S. Mattas, L. Saini and P. Jindal, "Secure Card-less ATM Transactions," 2020 First IEEE International Conference on Measurement, Instrumentation, Control and Automation (ICMICA), Kurukshetra, India, 2020, pp.1–4.
14. S Korouhanba Khuman, Gayatri Kalita, Basab Nath, "Card-less Automated Teller Machine: A New Simple Approach," International Journal of Scientific Research in Computer Science and Engineering, Vol.9, Issue.2, pp.27–31, 2021.

Index

A. Haldorai et al. (eds.), *6th EAI International Conference on Big Data Innovation for Sustainable Cognitive Computing*, EAI/Springer Innovations in Communication and Computing, https://doi.org/10.1007/978-3-031-54696-9

SPRINGER NATURE

GPSR Compliance

The European Union's (EU) General Product Safety Regulation (GPSR) is a set of rules that requires consumer products to be safe and our obligations to ensure this.

If you have any concerns about our products, you can contact us on ProductSafety@springernature.com

In case Publisher is established outside the EU, the EU authorized representative is:

Springer Nature Customer Service Center GmbH
Europaplatz 3
69115 Heidelberg, Germany

The manufacturer's authorised representative in the EU is Springer Nature Customer Service Centre GmbH, Europaplatz 3, 69115 Heidelberg, Germany. If you have any concerns regarding our products, please contact ProductSafety@springernature.com

Printed and bound by CPI Group (UK) Ltd, Croydon, CR0 4YY
01/10/2026
02225934-0002